D1518572

HARDWARE/SOFTWARE CO-DESIGN
FOR DATA FLOW DOMINATED
EMBEDDED SYSTEMS

HARDWARE/SOFTWARE CO-DESIGN FOR DATA FLOW DOMINATED EMBEDDED SYSTEMS

by

RALF NIEMANN
University of Dortmund,
Department of Computer Science XII,
Dortmund, Germany

KLUWER ACADEMIC PUBLISHERS
BOSTON / DORDRECHT / LONDON

A C.I.P. Catalogue record for this book is available from the Library of Congress.

ISBN 0-7923-8299-4

Published by Kluwer Academic Publishers,
P.O. Box 17, 3300 AA Dordrecht, The Netherlands.

Sold and distributed in North, Central and South America
by Kluwer Academic Publishers,
101 Philip Drive, Norwell, MA 02061, U.S.A.

In all other countries, sold and distributed
by Kluwer Academic Publishers,
P.O. Box 322, 3300 AH Dordrecht, The Netherlands.

Printed on acid-free paper

Printed in the Netherlands

Contents

List of Figures ix

List of Tables xiii

1. INTRODUCTION 1
   - 1.1 Embedded Systems 4
   - 1.2 Hardware/Software Co-Design 8
   - 1.3 Outline of the Thesis 11

2. CO-DESIGN SYSTEMS 13
   - 2.1 Co-Design for System Specification and Modelling 14
   - 2.2 Co-Design for Heterogeneous Implementation 14
     - 2.2.1 Processor Synthesis 14
     - 2.2.2 Single-Processor Architectures with one ASIC 15
     - 2.2.3 Single-Processor Architectures with many ASICs 16
     - 2.2.4 Multi-Processor Architectures 18
     - 2.2.5 Other Co-Design Approaches 20
   - 2.3 Comparison of Co-Design Approaches 20
   - 2.4 Overview of COOL 22
   - 2.5 Classification of COOL 24

3. SPECIFICATION OF EMBEDDED SYSTEMS 27
   - 3.1 Models of Computation 28
     - 3.1.1 State-Oriented Models 28
     - 3.1.2 Activity-Oriented Models 30
     - 3.1.3 Structure-Oriented Models 30
     - 3.1.4 Data-Oriented Models 31
     - 3.1.5 Heterogeneous Models 32

3.2 Requirements for Embedded System Specification 33
3.3 Survey of System Specification Languages 36
3.4 VHDL for System Specification 40
3.5 System Specification in COOL 41

4. HARDWARE/SOFTWARE PARTITIONING 47
4.1 The Hardware/Software Partitioning Problem 49
4.1.1 Hardware/Software Mapping 50
4.1.2 Hardware Sharing 51
4.1.3 Interfacing 52
4.1.4 Scheduling 52
4.1.5 Functional Pipelining 52
4.2 Related Work 53
4.3 Hardware/Software Partitioning in COOL 55
4.3.1 Special Aspects of HW/SW Partitioning in COOL 58
4.4 Hardware/Software Cost Estimation 62
4.4.1 Software Cost Estimation 64
4.4.2 Hardware Cost Estimation 67
4.4.3 Interface Cost Estimation 70
4.5 Generation of the Partitioning Graph 71
4.6 Formulation of the HW/SW Partitioning Problem 74
4.7 Optimization 80
4.7.1 Linear and Integer Linear Programming 81
4.7.2 Evolutionary Algorithms 83
4.8 The MILP-Model 86
4.8.1 The Decision Variables 89
4.8.2 The Objective Function 91
4.8.3 The Constraints 91
4.8.4 Hardware Sharing 94
4.8.5 Interfacing 95
4.8.6 Scheduling 97
4.9 HW/SW Partitioning based on Heuristic Scheduling 101
4.9.1 Step I: Mapping with Approximated Schedule 103
4.9.2 Step II: Scheduling solved by MILP 106
4.9.3 Step II: Scheduling solved by List Scheduling 106
4.9.4 Results I: Scheduling Algorithms 110
4.9.5 Results II: HW/SW Partitioning using Heuristic Scheduling 113
4.10 HW/SW Partitioning based on Genetic Algorithms 115

4.10.1 GA Parameters 116
4.10.2 GA Encoding 117
4.10.3 GA Fitness Function 119
4.10.4 Analysis of the GA Approach 123
4.10.5 Results III: HW/SW Partitioning based on GA 123
4.10.6 GA Approach for Extended HW/SW Partitioning 127

5. HARDWARE/SOFTWARE CO-SYNTHESIS 133
5.1 The Co-Synthesis Problem 134
5.1.1 Communication Synthesis 134
5.1.2 Specification Refinement 136
5.2 Related Work 143
5.3 Co-Synthesis in COOL 144
5.4 Hardware/Software Systems generated by COOL 146
5.4.1 System Controller 148
5.4.2 I/O Controller 151
5.4.3 Bus Arbiter 153
5.4.4 Controller for Data Paths 154
5.4.5 Bus Drivers 154
5.4.6 Simulation Models for Memory and Processors 155
5.5 State-Transition Graph 156
5.5.1 State-Transition Graph Generation 157
5.5.2 State-Transition Graph Optimization 161
5.5.3 Memory Allocation 163
5.6 Refinement and Controller Generation 164
5.6.1 Data-Related Refinements for Hardware and Software 164
5.6.2 Software Refinement 167
5.6.3 Hardware Refinement 170
5.6.4 Controller Generation 171
5.7 Results IV: Application Study of a Fuzzy Controller 174

6. THE COOL FRAMEWORK 183
6.1 Implementation of COOL 183
6.2 Description of COOL 185
6.2.1 Graphical User Interface 185
6.2.2 Validation using Simulation 186
6.2.3 Specification of Design Constraints 188
6.2.4 Hardware/Software Partitioning 189
6.2.5 Co-Synthesis 190

6.2.6 Co-Simulation 191

7. SUMMARY AND CONCLUSIONS 193
7.1 Summary and Contribution to Research 193
7.2 Future Work 196

References 199

Notations 213

Abbreviations 215

Index 217

List of Figures

1.1	The Y-chart	2
1.2	Heterogeneous system	7
1.3	Hardware/software co-design	9
2.1	Classification of co-design approaches	13
2.2	Single-processor architecture with one ASIC	15
2.3	Single-processor architecture with many ASICs	16
2.4	Multi-processor architecture	18
2.5	Categorization by specification power	21
2.6	Categorization by implementation power	22
2.7	Overview of COOL	23
2.8	COOL categorized by specification power	25
2.9	COOL categorized by implementation power	25
3.1	Petri net	28
3.2	FSM	29
3.3	Hierarchical concurrent FSM (HCFSM)	29
3.4	Data flow graph (DFG)	30
3.5	Control flow graph (CFG)	30
3.6	Block diagram (a), RT netlist (b) and gate netlist (c)	31
3.7	Entity relationship model	31
3.8	Control/data flow graph (CDFG)	32
3.9	Program state machine (PSM)	33
3.10	Specification languages for embedded systems	39
3.11	Behavioral (a) and structural (b) description in VHDL	40
3.12	System specification using COOL	42
3.13	System simulation in COOL	44

3.14	Design constraints specified with COOL	45
4.1	Hardware/software partitioning	49
4.2	Hardware/software mapping	50
4.3	Hardware sharing	51
4.4	Interfacing	52
4.5	Functional pipelining	53
4.6	Hardware/software partitioning in COOL	56
4.7	Communication modelling in COOL	58
4.8	Communication mechanisms	59
4.9	No computation during channel access	60
4.10	Violation of execution order	61
4.11	Cost estimation flow in COOL	63
4.12	Transformation of an unbounded loop	64
4.13	Computation flow for software cost estimation	65
4.14	Software execution time estimation	66
4.15	Design flow for hardware cost estimation	67
4.16	Function description	68
4.17	Transformation rule	68
4.18	Component specification	69
4.19	Control step list	69
4.20	Generation of the partitioning graph	72
4.21	Partitioning graph	74
4.22	Target technology	75
4.23	Hardware/software implementations	77
4.24	General evolutionary algorithm	84
4.25	Genetic algorithm encoding	84
4.26	Crossover and mutation	85
4.27	Node sets	87
4.28	Usage of decision variables	90
4.29	Constraints caused by target architecture restrictions	92
4.30	Hardware sharing	94
4.31	Interfacing	95
4.32	Application of scheduling constraints	97
4.33	Simplified scheduling model	98
4.34	Precise scheduling model	100
4.35	Number of variables for generated MILP models	101

4.36 Heuristic scheduling 102
4.37 Using predecessor nodes in heuristic scheduling 103
4.38 Using dominator nodes in heuristic scheduling 105
4.39 Additional edges 108
4.40 List scheduling approach 109
4.41 Computation of an AT-curve 111
4.42 Scheduling results 112
4.43 Scheduling precision 112
4.44 Computation time of scheduling algorithms 113
4.45 Deviation of computed hardware area 114
4.46 Computation time of heuristic scheduling partitioning algorithm 115
4.47 GA approach 116
4.48 Encoding the hardware/software partitioning problem with GA 118
4.49 Matrices to compute resource costs in linear time 120
4.50 Average relative deviation to best solution 124
4.51 Influence of crossover and mutation rates 125
4.52 Chip area obtained by GA partitioning algorithm 125
4.53 Deviation of computed hardware area for different partitioning algorithms 126
4.54 Computation time of partitioning algorithms 126
4.55 GA encoding for extended partitioning 128
4.56 Hardware implementation alternatives in extended partitioning 129
4.57 Alternative hardware implementations for components of `al4` 130
4.58 AT-curves of partitioning `al4` with different cost models 131
5.1 Overview of co-synthesis 134
5.2 Tasks of co-synthesis 135
5.3 Communication synthesis 135
5.4 Specification refinement 136
5.5 Control-related refinement 138
5.6 Little and big endian 139
5.7 Refinement of accesses to external data 140
5.8 Communication channels 141
5.9 Co-synthesis in COOL 145
5.10 Hardware/software system generated by COOL 147
5.11 Synchronization scheme I for processors 148
5.12 Synchronization scheme II for processors 150
5.13 Synchronization scheme V for ASICs 151

5.14 I/O controller 152
5.15 Bus arbiter 153
5.16 Data path controller 154
5.17 Bus driver 155
5.18 Local memory 156
5.19 States and transitions 157
5.20 State-transition graph (STG) generation 160
5.21 Transition elimination for the STG 161
5.22 State elimination for the STG 162
5.23 Example of STG optimization 163
5.24 Memory allocation 163
5.25 Operation replacement 166
5.26 Software refinement in COOL 168
5.27 Target-specific operations and refined software 169
5.28 Hardware refinement 170
5.29 Transformation of arrays into local memory accesses 171
5.30 Generation of controller description 172
5.31 Memory access by hardware components 173
5.32 Crossing situation 175
5.33 Block diagram of the fuzzy controller 175
5.34 COOL specification of the fuzzy controller 176
5.35 Heterogeneous target architecture for the fuzzy controller 177
5.36 Prototyping environment to implement the fuzzy controller 177
5.37 Prototyping board 178
5.38 Hardware/software partitioning of the fuzzy controller specification 179
5.39 Hardware/software netlist of the fuzzy controller 180
5.40 Simulating the netlist of the fuzzy controller 182
6.1 Graphical user interface of COOL 186
6.2 Simulation of Specification with VANTAGE OPTIUM 187
6.3 Specification of design constraints 188
6.4 Design space exploration in Hardware/software partitioning 189
6.5 Design space exploration in co-synthesis 190
6.6 Simulation of system specification and netlist 191

List of Tables

1.1 Design objects on different levels of abstraction 2
4.1 Comparison of different partitioning approaches 55
4.2 Classification of hw/sw partitioning approaches in COOL 58
4.3 Overview of evolutionary algorithms 85
4.4 Set of benchmarks 110
4.5 Results of scheduling algorithms 113
4.6 Results of partitioning algorithms 114
4.7 Comparison of P_HS and P_GA for hardware/software partitioning 127
5.1 Communication channels 141
5.2 Synchronization schemes supported by COOL 151
5.3 Data type conversion for system specification 165
5.4 Examples of operation replacement 166
5.5 Folding of signals into memory accesses 167

Foreword

Many of the modern applications of microelectronics require huge amounts of computations. Despite all recent improvements in fabrication technologies, some of these computations have to be performed in hardware in order to meet deadlines. However, controlling computations by software is frequently preferred due to the larger flexibility. Hence, in general, modern applications require a mix of software-based and hardware-based computations. Applications using this mix can be designed with the help of hardware/software co-design systems. Many such co-design systems have been described so far (references can be found in this book), but many of these are based on heuristics. In this book, Niemann describes a co-design system which is based on sound modeling techniques. This system has the following salient features:

- *Precise cost and performance figures*
  Design decisions for implementing a certain function in hardware or software are based on cost and performance figures for the different design alternatives. Hence, good designs can only be expected if these figures are accurate. In order to achieve excellent accuracy, Niemann takes a new approach:

  - the cost of software implementations is derived from the data available about the target processors and from knowledge about the code size.

  - the performance of software implementations is computed by compiling the given function and then using static analysis for computing worst case execution times.

  - the cost of hardware implementation is estimated by running higher-level synthesis tools.

  - the performance of hardware implementations is again computed by using static analysis.

This approach requires some (but not excessive amounts of) time for computing these figures. A benefit of this approach consists of the small number of design iterations.

- *Precise partitioning*
  Partitioning is the task of selecting either a software or a hardware implementation for a certain function. Niemann describes a closed mathematical model of the partitioning problem. Due to a proper choice of the granularity at which Niemann describes functions, the mathematical problem can be solved in reasonable time if appropriate optimization methods are used.

- *Demonstration of the power of genetic algorithms*
  In this book, Niemann shows that genetic algorithms are appropriate for solving the partitioning problem. These algorithms solve the problem within predictable run-times and can handle even the extended partitioning problem. The extended partitioning problem integrates the selection of a suitable hardware implementation for each function into the partitioning process.

- *Completely integrated hardware/software co-design system*
  The method proposed by Niemann has been implemented in the hardware/ software co-design tool called COOL. COOL provides complete integration of all co-design steps, including specification capture, partitioning, cosimulation and synthesis of all required controllers.

- *Validation of* COOL *for real designs*
  COOL has been validated by generating real mixed hardware/software systems

I think that the method proposed by Niemann is a key contribution towards more precise models and better optimization techniques for hardware/ software co-design. Such techniques are required in order to generate efficient designs that are accepted by industry. Niemann's method should be known by all persons working in the field. Hence, I recommend this book for everyone who is interested in hardware/software co-design.

Dortmund, July 17th, 1998 Peter Marwedel

Preface

In the past years, design automation techniques and systems have been developed supporting the design process of embedded systems. Nowadays, high-level synthesis tools represent state-of-the-art tools to design a single ASIC from a behavioral description. The next goal in ECAD research is to raise the abstraction level from high-level synthesis to system-level synthesis supporting the design process of heterogeneous systems consisting of both ASICs and programmable processors. For this reason, hardware/software co-design has become a very important issue in ECAD research and industry. One key issue of hardware/software co-design is to exploit both the advantages of "software" (programmable processors) and "hardware" (ASICs). The usage of software in the design process of embedded systems leads to shortened product cycles, reduced design costs, improved maintenance, and ease of debugging and testing. Therefore, hardware/software co-design gains more and more importance.

Features of this book

This book addresses the increased importance of hardware/software co-design. The main co-design problems are introduced, including system specification, cost estimation, hardware/software partitioning and co-synthesis. Main emphasis is put on hardware/software partitioning, co-synthesis and their coupling. The book provides an overview of existing co-design techniques and systems. It is the first book which defines a precise mathematical model for the hardware/software partitioning problem supporting multi-processor-multi-ASIC target architectures. This model is suitable for data flow dominated systems. Different algorithms are compared to solve this mathematical model. A new concept is introduced for co-synthesis to automatically generate a set of hardware and software specifications prepared to implement the complete system by software compilation and hardware synthesis. The hardware/software co-design tool COOL is presented which supports the complete design flow from system specification down to hardware/software implementation. All techniques are described in detail and are exemplified. Therefore, the book is intended to serve academic researchers, but also designers and engineers from industry.

Acknowldegements

This book represents a revised version of my doctoral thesis submitted to the Department of Computer Science at the University of Dortmund in January 1998. My gratitude is to my advisor Prof. Dr. Peter Marwedel for giving me the opportunity and the freedom to perform research in the area of hardware/software co-design. Furthermore, I would like to thank my co-referee Prof. Dr. Heinrich Müller for his efforts.

Many thanks to all my colleagues at the 'LS XII' group of the Department of Computer Science at the University of Dortmund. In particular, I want to thank Steven Bashford, Dr. Birger Landwehr and Dr. Rainer Leupers for valuable discussions. In addition, I gratefully acknowledge the support of Prof. Dr. Anupam Basu, Heiko Falk, Christoph Fritsch, Beate Hendges, Petra und David Knowles, Hiroshi Shinkai and Dr. Karl-Heinz Temme.

This book is dedicated to my family, in particular my wife Sabine, our daughter Franziska and our son Alexander who was born two weeks ago. I am especially grateful to my parents, Marlies and Willi Niemann, because their support allowed me to study computer science. All of them gave me the love, patience and understanding to write this book.

Dortmund, July 14th, 1998 Ralf Niemann

1 INTRODUCTION

In the last two decades, the state of *computer-aided design* (CAD) of digital systems has changed substantially. The level of abstraction has always been raised to higher levels, when the design steps of lower levels had been supported by tools. As a consequence, circuits with over one million devices were manufactured successfully in the late 1980s. Nowadays, ECAD[1] vendors sell behavioral (or high-level) synthesis tools enabling designers to handle such large designs. The goal of the nineties has been to raise the level of abstraction to the so-called *system level.* On this level complex systems are not implemented by designing pieces (e.g. single chips) of the system separately, but by starting the design process from a single system-level specification.

The *Y-chart* [GaKu83] in figure 1.1 is used to describe the different views and abstraction levels of *VLSI*[2] systems. To implement a system, a designer iteratively refines the system specification by going down the abstraction levels. The overall goal is to implement the system at the *physical level* at which a circuit can be manufactured. The axes of the Y-chart have the following meaning: The *behavioral domain* defines the behavioral form of a design without any information of its implementation. In contrast, the *physical domain* describes the physical characteristics of a given target architecture with which the sys-

[1]ECAD : Electronic Computer Aided Design
[2]VLSI : Very Large Scale Integration

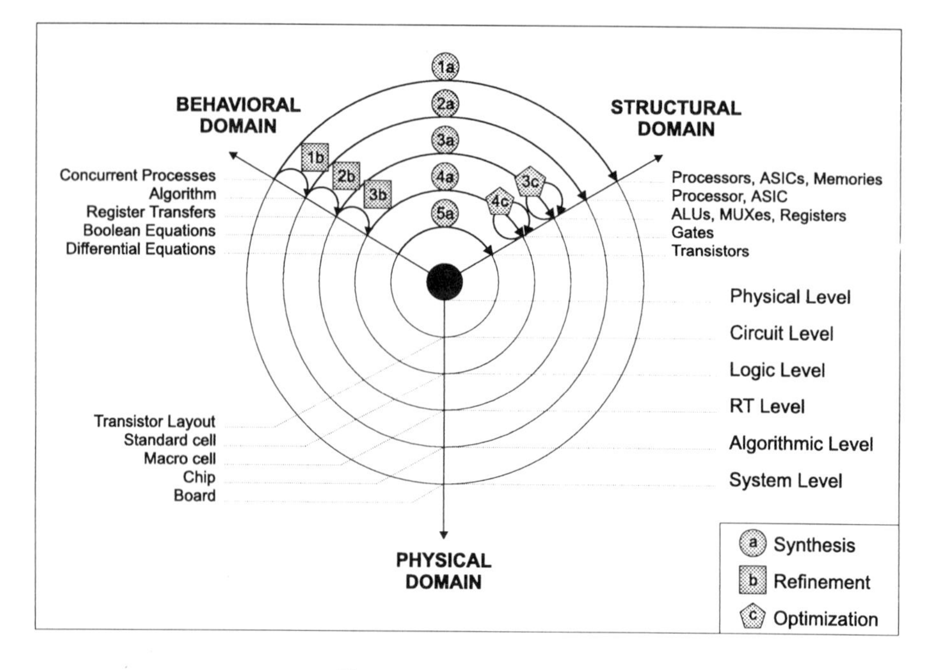

Figure 1.1. The Y-chart

tem should be implemented. The *structural domain* bridges the behavioral and physical domain. It describes the system as a set of interconnected functional components on which the behavioral description is mapped. Table 1.1 gives an overview of these three domains and their design objects.

	Level	Behavioral forms	Structural forms	Physical objects
1	system level	concurrent processes, communication	processors, controllers, ASICs, memories, busses	board, cabinets, cables
2	algorithmic level	algorithms, statements, loops, operations	processor, ASIC	block, chip
3	RT level	register transfers	registers, ALUs, MUXs	ICs, macro cells
4	logic level (gate level)	boolean equations, finite-state machines	logic gates, flip-flops	standard cells
5	circuit level (transistor level)	transistor functions, network equations	transistors, resistors, capacitors	transistor layout, analog cells, digital cells
6	physical level	diffusion behavior	chemical elem.	masks, polygons

Table 1.1. Design objects on different levels of abstraction

In the past years different design tasks have been identified including synthesis, refinement and optimization steps:

1. *Synthesis* steps represent the transition from the behavioral to the structural domain.

2. *Refinement* steps refine an abstract specification into a more detailed one.

3. *Optimization* steps maximize the quality of a design, e.g. by minimizing its cost, while staying in the same domain.

Using these terms, the different abstraction levels can be described as follows: At the lowest level, *physical design* creates the physical layout of a chip. The main tasks in physical design are *placement* and *routing*. The design at the *circuit level* includes *circuit-level synthesis* (5a in figure 1.1) transforming a set of transistor equations into a transistor schematic including transistor types, parameters and sizes. At the *logic level*, synthesis and optimization steps are applied. *Logic-level synthesis* (4a) is used to generate the structural view of a *logic level* model, e.g. boolean equations. *Logic-level minimization* (4c) minimizes the hardware area. The result of logic-level design is a minimized *gatelevel* netlist of the design which is finally bound to cells of a given library, known as *library binding* or *technology mapping*. The main task at the *register-transfer level* is *controller synthesis* (3a). The controller descriptions are refined (3b) into a set of interacting finite-state machines at the logic level. Typical optimization steps (3c) at the RT-level are *state minimization*, *state encoding* and minimization of the related combinational components. At the *algorithmic level*, *behavioral* or *high-level synthesis* (2a) transforms the behavioral description of an algorithm into a structural representation at the register-transfer level including registers, arithmetic units and multiplexers. It computes a *data path* containing hardware resources and a *data path controller* steering the data path. Typical synthesis steps (2a) at the algorithmic level are the following: *allocation* of hardware resources being able to implement an operation of the algorithm, *scheduling* the execution times of these operations, and *binding* them to resources.

All design tasks described before gained industrial acceptance during the past years and a variety of commercial tools are available nowadays for them. But all these synthesis tasks are concerned with the design of a single hardware chip. The next goal in ECAD research is the design of complex systems starting from a single system-level specification. The overall goal is to implement systems on a target architecture at the board or chip level. In addition, the target architecture may be heterogeneous, containing processors, ASICs[3], memories and busses. This research area in ECAD is called *system-level design*. The task of system-level design is to implement (1a) a system-level specification on

[3] ASIC : Application Specific Integrated Circuit

a target architecture by refining (1b) the implementation-independent system-level specification into a set of target-specific specifications.

1.1 Embedded Systems

There are two different classes of digital systems: general-purpose systems and special-purpose systems. *General-purpose systems* are, for example, traditional computers ranging from PCs, laptops, workstations to super-computers. All these computers can be programmed by a user and support a variety of different applications determined by the executed software. General-purpose systems are not developed for special, but for general applications. During the eighties and early nineties, these general-purpose systems were the main drivers for the evolution of design methods. This situation is changing, because several new applications (e.g. MPEG[4]) have come up which cannot be implemented efficiently by general-purpose systems. For this reason, *special-purpose systems* gain more and more importance, because they are developed to fulfill a special, fixed task. Most of these systems are configured once and work independently afterwards. The user has limited access to program these systems. Special purpose systems can be divided into terminal-based and *embedded systems* which can be characterized by the following features: Embedded systems ...

1. are usually embedded in other products.
2. are working stand-alone.
3. are infrequently reprogrammed. Their functionality is mostly fixed.
4. work very often in a *reactive* mode, responding frequently to external inputs.
5. are implemented by numerous concurrently working processes requiring *inter-process communication.*
6. have stringent time requirements, e.g. *real-time constraints.*
7. are very often I/O intensive.
8. are extremely sensitive concerning cost-, power-, and performance-criteria.
9. have hard reliability and correctness constraints, e.g. an *ABS*[5] system has to work bug-free in any situation.

The importance of embedded systems is highlighted by the fact that in 1995 43% of sold 32-bit processors were used in computing applications, the remaining 57% for embedded systems [PGL+97]. Furthermore, 8- and 16-bit processors dominate the embedded system market with 95% in volume, in contrast to general-purpose systems. The design process of embedded systems is always

---

[4]MPEG : Motion Picture Expert Group

[5]ABS : Anti-lock Brake System

driven by the following goal: *"Maximize the system value while minimizing the cost"*. The value and the cost of a system are always dependent on each other and can be measured by the following criteria:

1. *time-to-market*, short product cycles (e.g. important for consumer products),
2. *performance* (e.g. to guarantee real-time constraints),
3. *power consumption* (important for portable applications),
4. *form factors* (small weight and physical volume are important for portable applications),
5. ease of *debugging* and *testing*,
6. ease of *programmability* (e.g. important for new generation of products),
7. *safety* and *reliability* (e.g. important for control-systems),
8. *maintenance* and *servicing* issues,
9. *design costs* (e.g. number of hardware parts that need to be designed, or required software compilers),
10. *manufacturing costs* (e.g. size of silicon, packaging costs).

The optimal design process from a management point of view is the following: *"Produce more flexible products in shorter time to market. Additionally, design time, development and production cost have to be predictable."*

Embedded systems can also be divided into two different subclasses: embedded controllers and embedded data-processing systems.

Embedded controllers are dedicated to control functions. They are *control flow dominated* and react to external events. Therefore, embedded controllers are very often called *reactive systems*. Reactive systems typically respond to incoming stimuli from the environment by changing its internal state and producing output results. Normally, they support a set of modes and settings and their real-time constraints are often in the range of milliseconds. Therefore, the performance required usually varies from low to moderate. For this reason, *microcontrollers* are very often sufficient to implement an embedded controller. Typical applications of embedded controllers are the following:

1. controllers (e.g. elevators, electronic window-lifters),
2. home appliances (e.g. microwave ovens, washing machine),
3. automobile industry applications (e.g. engine control unit, fuel injection, anti-locking brakes),
4. industrial robots.

Embedded data-processing systems are dedicated to data communication and processing. Therefore, they are very often called *transformational systems*. These systems are *data flow dominated* and often they are *real-time systems* executing a special function within a predefined time window. They require a much higher performance compared to embedded controllers. Therefore, microcontrollers are not sufficient and more powerful microprocessors (very often DSPs[6] or ASIPs[7]) and ASICs are required. In most cases, the model of computation is a *synchronous* or *dynamic data flow model* of computation. This is the classical domain of *digital signal processing* and *high-level synthesis*. The following application domains contain typical examples of embedded data-processing systems:

1. multi-media:
   - audio applications:
     - Dolby AC3,
     - MPEG2 that support multi-channel surround audio,
   - video applications:
     - MPEG2 decoders for set-top boxes for satellite and cable digital TV,
     - video coding standards from JPEG[8], to MPEG1, MPEG2 and eventually to MPEG4,
     - digital video disks (DVD),
     - high-definition TV (HDTV),
     - digital video broadcast (DVB),
     - video-phone,
     - 3-dimension video,
2. consumer electronics:
   - video games,
   - office requirements (printers, fax-machines),
3. wireless communication:
   - GSM[9] digital cellular,
   - DECT[10] cord-less telephone,

[6]DSP : Digital Signal Processor
[7]ASIP : Application Specific Instruction-Set Processor
[8]JPEG : Joint Photographic Expert Group
[9]GSM : Global System for Mobile Communication
[10]DECT : Digital European Cord-less Telephone

- North American digital cellular standards such as IS-54B digital cellular,

4. general telecommunication:
   - automatic call distribution and operator headset voice processing,
   - ATM[11],
   - low power wireless base stations for cord-less telephone standards CT2 and CT2+,
   - ISDN[12] phone terminals,
   - fax and modems,
   - digital answering machines.

These embedded data-processing systems can also be divided into two subclasses where the required performance is the main difference. *Audio* and *telecommunication* systems have *sample rates* ranging from 10^4 to 10^6 samples per second, in contrast to typical *video* algorithms which have sample rates ranging from 10^7 to 10^8 samples per second. These algorithms require 1-10 billion operations per second, and a bandwidth of more than 1 Gbyte/s. Embedded systems are often implemented by *heterogeneous systems* consisting of dedicated and programmable parts. Therefore, they are well-known as *hardware/software systems* where "hardware" represents the dedicated hardware parts, and "software" the programmable processors. A typical heterogeneous system is depicted in figure 1.2.

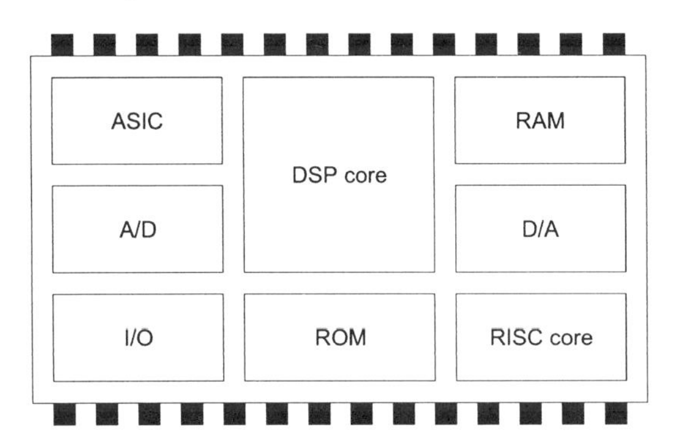

Figure 1.2. Heterogeneous system

Heterogeneous systems contain dedicated hardware parts (ASICs) and programmable embedded processors. These processors may be, for example, DSP or RISC[13] processors. In addition, RAM is required for storing data and ROM

[11] ATM : Asynchronous Transfer Mode and data networks
[12] ISDN: Integrated Service Digital Network
[13] RISC : Reduced Instruction-Set Computer

for program code and constant tables. Peripherals (A/D converter, D/A converter and I/O unit) are necessary for communicating with the environment. A network of busses and wires connects all these components. Heterogeneous systems can be integrated on a single board or an a single chip. *Single-board systems* are integrated on a board containing components, like ASICs, processors, ASIPs and memories. *Single-chip systems* are integrated on one ASIC containing processor cores, dedicated parts and memories. The processor cores of a single-chip system are provided in the component library of the ASIC technology. The advantages of single-chip solutions compared to single-board solutions are the increased performance and reliability. In addition, the power consumption and the manufacturing costs are reduced. On the other hand, single-chip systems have a larger chip size and therefore debugging those systems becomes much harder. The design problems of both integration level categories are comparable and will be described next.

1.2 Hardware/Software Co-Design

Hardware/software co-design can be defined as the cooperative design of hardware and software. Co-design research deals with the problem of designing heterogeneous systems. One of the goals of co-design is to shorten the time-to-market while reducing the design effort and costs of the designed products. Therefore, the designer has to exploit the advantages of the heterogeneity of the target architecture. The advantages of using processors are manifold, because software is more flexible and cheaper than hardware. This flexibility of software allows late design changes and simplified debugging opportunities. Furthermore, the possibility of reusing software by porting it to other processors, reduces the time-to-market and the design effort. Finally, in most cases the use of processors is very cheap compared to the development costs of ASICs, because processors are often produced in high-volume, leading to a significant price reduction. However, hardware is always used by the designer, when processors are not able to meet the required performance. This trade-off between hardware and software illustrates the optimization aspect of the co-design problem. Co-design is an interdisciplinary activity, bringing concepts and ideas from different disciplines together, e.g. system-level modelling, hardware design and software design.

The design flow of the general co-design approach is depicted in figure 1.3. The co-design process starts with specifying the system behavior at the system level. After this, the *system specification* is divided into a set of smaller pieces, so-called *granules* (e.g. basic blocks). In a *cost estimation* step, values for some cost metrics are determined for these granules. These cost metrics include estimations for hardware or software implementations. Hardware cost metrics are, for example, execution time, chip area, power consumption or testability. Software cost metrics may include execution time and the amount of required program and data memory. After the cost estimation has been

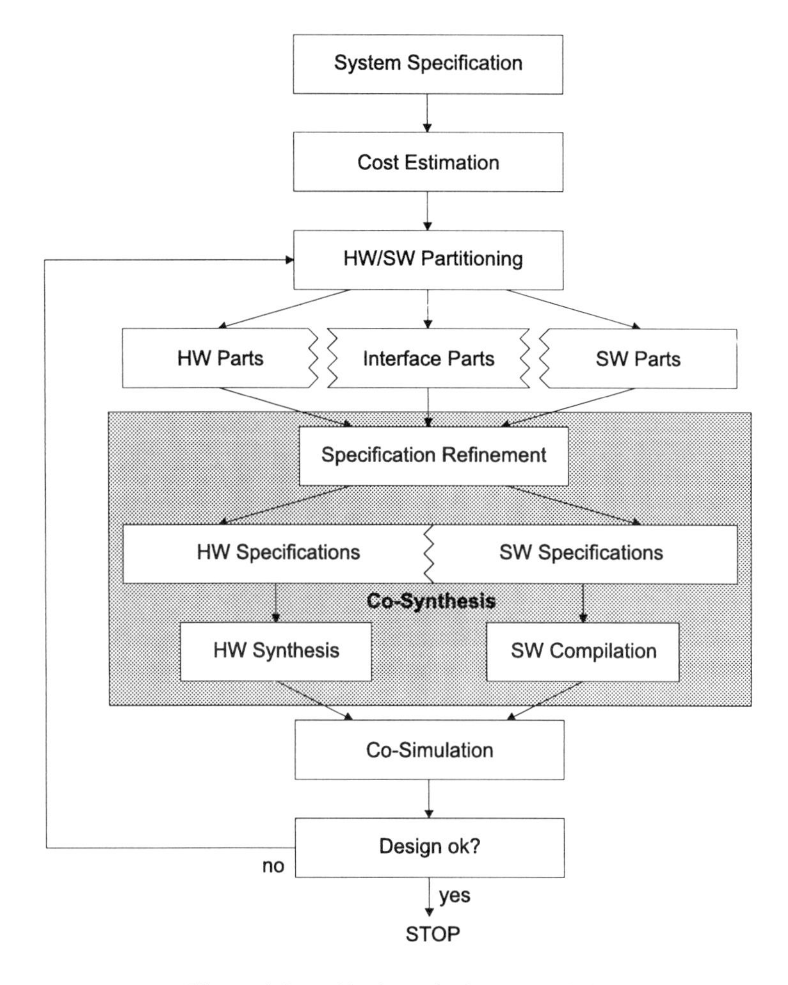

Figure 1.3. Hardware/software co-design

performed, the *hardware/software partitioning* phase computes a good mapping of these granules to hardware or software resulting in sets of granules implemented on hardware (hardware parts) or software (software parts). To implement the system on a heterogeneous target architecture, the mapping requires additional interface parts (implementing the communication and synchronization) between ASICs and processors. The *specification refinement* step transforms the implementation-independent system specification into hardware and software specifications. All specifications include communication mechanisms to allow the exchange of data between processors and ASICs. Hardware is synthesized from the given specification, the software specification is compiled for the chosen processor. The result of this *co-synthesis* phase is a set of ASICs and a set of assembler programs for the processors. In a final *co-simulation* step, the ASICs are simulated together with the processors executing their generated assembler programs. If all performance constraints are met and the cost of the design is acceptable, the co-design process stops, otherwise a *re-partitioning* step is executed to optimize the design until a sufficient system implementation has been found.

In addition to the presented problems, there are further problems in the area of hardware/software co-design:

- The *co-validation* problem in system-level design includes different methods to detect errors at different abstraction levels. Co-validation methods include formal verification, simulation or emulation.

  - *Formal verification* allows to prove formally either the equivalence of different design representations or specific properties, e.g. the absence of *dead-lock* conditions, of the system specification. Therefore, formal verification represents an important issue in particular for *safety-critical* applications, e.g. ABS. Formal verification of hardware/software systems is often referred as the *co-verification* problem.

  - *Simulation* validates the functional correctness for a set of input stimuli. In most cases, only a small set of all combinations of input stimuli can be simulated. For this reason, simulation only ensures the correct behavior with a certain probability. Simulation can be applied during different design steps including also the co-simulation step after co-synthesis as described before.

  - To speed up the simulation time for simulating a partitioned hardware/software system, *emulation* is used. Emulation systems map the ASICs onto programmable hardware, e.g. FPGA[14] and couple them with processors on a board. Therefore, emulators provide the closest to real prototypes that is possible.

- Another important related problem in co-design is *software compilation*. In contrast to general-purpose systems, the *quality* of the generated assembly code has more importance than the *compilation time*. State-of-the-art DSP-compilers, for example, produce an overhead of 200-500% (in some cases even more) as compared to hand-crafted assembler code [ZVSM94]. Therefore, most software development for DSPs is done manually now a days. For this reason, an important goal is the development of (DSP-)specific code-optimizations for compilers to improve the code quality. A good overview of code generation techniques for embedded processors is given in [MaGo95]. Very often, ASIPs are used to implement embedded systems. In most cases, no compiler exists for these ASIPs. Therefore, *retargetable compilers* [Leup97], [Liem97] have gained great importance to generate code for ASIPs automatically.

[14]FPGA: Field Programmable Gate Array

- A problem related to hardware/software partitioning is *design space exploration*, where the partitioning algorithm should produce a number of different solutions in short computation time. This enables the designer to compare different design alternatives to find appropriate solutions for different objective functions, e.g. high-performance, low-cost or low-power designs.
- A unified *design methodology management* supporting specification, validation and co-synthesis of both hardware and software is the overall goal of the co-design research.

1.3 Outline of the Thesis

Hardware/software co-design has become a very important research area in VLSI design. The overall goal is to develop tools supporting the concurrent design of hardware and software. This book presents new approaches for some phases of co-design targeted to data flow dominated applications. These new approaches are not only theoretical, but have been integrated in a complete hardware/software co-design framework (COOL[15]). This framework is on top of state-of-the-art tools for hardware (high-level synthesis) and software design (standard, as also retargetable compilers). The correct functionality of COOL has been validated through several application studies. Main emphasis of this co-design approach is put on heterogeneous implementation, including hardware/software partitioning and co-synthesis.

The organization of the book is as follows: Chapter 2 presents existing co-design approaches and categorizes them. A short overview and classification of COOL is given. The following chapters will describe the different co-design steps, including an overview of related work and their solutions in COOL. Chapter 3 discusses different system specification models required for embedded systems and compares specification languages used in practice. In addition, the method of specifying systems with COOL is presented. In chapter 4, an overview of existing hardware/software partitioning approaches is given. Then, three new algorithms are presented for solving the partitioning problem. These are based on mixed integer linear programming (MILP), an iterative approach combining an MILP formulation with heuristic, and a genetic algorithm approach. This variety of algorithms allows the designer to compute high-quality results and/or to explore the design space. In chapter 5, the co-synthesis step for a partitioned system is described. The refinement of the system specification, including memory allocation, generation of a run-time scheduler, interface and communication synthesis will be illustrated. The hardware/software co-design framework COOL will be presented in chapter 6. Finally, a conclusion and an overview of future extensions is given in chapter 7.

---

[15]COOL: Hardware/Software Co-design Tool

2 CO-DESIGN SYSTEMS

Several research groups in the area of hardware/software co-design have developed tools supporting some aspects of co-design problems. In this chapter an overview and a classification of existing work will be given.

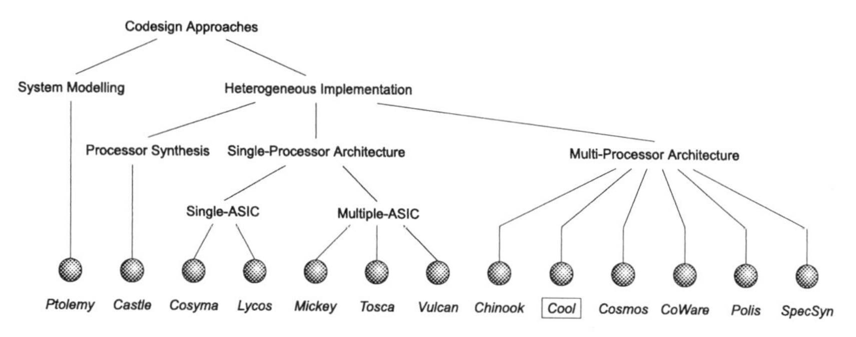

Figure 2.1. Classification of co-design approaches

In figure 2.1, some of the main co-design approaches are classified based on their key research issue. A small number of research groups emphasize the problem of *system specification and modelling*, but most of them work on *heterogeneous implementation*. This research area includes mainly the hardware/software partitioning and co-synthesis problem. The complexity of the supported target architecture is one main criterion to distinguish the existing approaches.

First, different co-design systems mainly focussing on system specification will be described in section 2.1. Then, co-design tools for heterogeneous implementation are presented in section 2.2. In section 2.3 these co-design systems will be compared concerning their specification and implementation power. Finally, the co-design tool COOL will be introduced in section 2.4 and classified in section 2.5.

2.1 Co-Design for System Specification and Modelling

Ptolemy. PTOLEMY [BHLM91, KaLe93] is a framework for the specification, simulation, prototyping and software synthesis of digital signal processing systems. It is an object-oriented framework allowing to specify complex systems by mixing different computational models. Each model of computation is called a *domain*. The possibility of mixing different domains allows the designer to specify parts of the system in the most natural and efficient manner. Examples of domains are *synchronous data flow* (SDF), *dynamic data flow* (DDF) and *discrete event* (DE). Each domain is specified in a procedural C++-type language and consists of a set of stars and galaxies. A *star* is a primitive of a domain, e.g. data flow operators or logic gates. A *galaxy* is a collection of stars and other galaxies. A *scheduler* is included within each domain to determine the execution order of the stars. Whenever different domains are mixed, a *worm-hole mechanism* is created facilitating the communication between the two schedulers with the appropriate data type conversion. A negative consequence of the strength of a heterogeneous system specification is the loss of analytical power. But nevertheless, PTOLEMY represents an important step towards heterogeneous specification and simulation of complex systems.

Others. There are some other tools working on system specification of complex system. The KHOROS [KoRa94] tool has been developed for rapid prototyping of image and signal processing algorithms. In the area of reactive systems, some commercial variants of STATEMATE [HaPn88] have been established.

2.2 Co-Design for Heterogeneous Implementation

2.2.1 Processor Synthesis

Castle. CASTLE[1] [CaWi96, WiCa97] is a co-design workbench assisting the designer to find a cost-effective implementation of a system. The co-synthesis approach of CASTLE differs from the traditional one. The target architecture

[1] CASTLE: Codesign and Synthesis Tool Environment

is not fixed, but the goal of CASTLE is to synthesize a processor and a program for this processor implementing the system behavior. First, the designer specifies the system with a common specification language, e.g. C++, VHDL or Verilog. Front-ends compile this specification into the SIR[2] format based on control/data flow graphs. SIR is the basic intermediate format on which a variety of tools in CASTLE are working. Analysis tools supply the design with *static* or *dynamical profiling* information of the specified algorithm. Based on this information, the designer specifies the main structure of the processor using the schematic entry of CASTLE. The processor is specified by using generic VHDL descriptions of an architecture library containing processor components like adders, registers, etc. Then, the co-synthesis step in CASTLE generates a synthesizable VHDL description of the processor and a corresponding compiler back-end. These design steps can be iterated until the designer is satisfied with the partitioning of the system and the resulting performance.

2.2.2 Single-Processor Architectures with one ASIC

Co-design tools supporting *single-processor architectures* try to speed-up the total execution time of a program running on the processor, by adding an ASIC as a *co-processor* to the target architecture. The typical target architecture is depicted in figure 2.2.

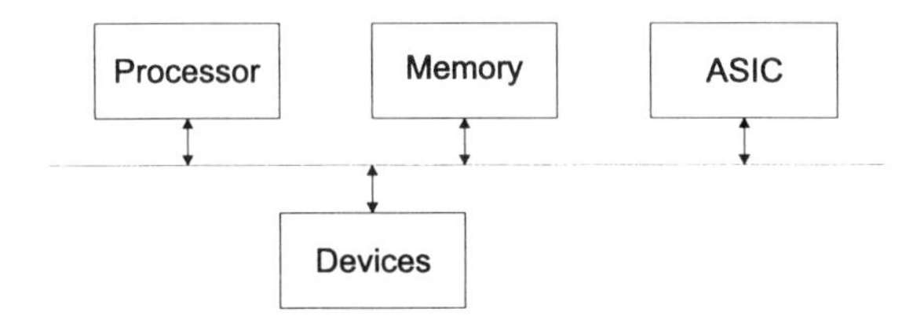

Figure 2.2. Single-processor architecture with one ASIC

COSYMA and LYCOS are typical representatives of these approaches. The main difference between both approaches is that LYCOS supports concurrency between processor and ASIC which COSYMA does not.

Cosyma. COSYMA[3] [EHB93, HEY+95] is a co-design system focussing on a co-processor target architecture as depicted in figure 2.2 where the processor is a *RISC* CPU. The goal of COSYMA is to compute a hardware/software partition of a specified system to obtain a maximized speed-up by using the co-processor. Systems are specified in COSYMA using the C-like specification language C^x which allows to define concurrency and timing constraints. Then,

[2]SIR : System Intermediate Representation
[3]COSYMA : Co-synthesis for Embedded Architectures

the C^x specification is compiled into an internal *extended syntax graph* representation which is used for analysis and simulation to get *profiling* information. Partitioning is done automatically based on *simulated annealing* using estimated schedule times. Finally, the resulting hardware parts are synthesized with the high-level synthesis tool BSS[4]. The main limitation of COSYMA is that the processor and the co-processor can not work in parallel. As a consequence, possible concurrency is not exploited.

Lycos. LYCOS[5] [MGK+97] is a co-design tool with emphasis on *design space exploration* using automatic hardware/software partitioning. The target architecture is also a single-processor architecture with one ASIC as a co-processor. LYCOS uses an implementation independent intermediate format called *Quenya* which is based on communicating control/data flow graphs. A variety of tools are built around Quenya. The designer specifies the system in VHDL or C and translates it into a Quenya control/data flow graph. Then, a *profiling* step is performed to get some execution statistics required to find bottlenecks in the algorithm. Before partitioning, the target architecture has to be selected containing a processor for the software and a hardware technology for manufacturing the ASIC. The number of available hardware modules implemented on the ASIC can be constrained by the user. The communication scheme between processor and ASIC is fixed to *memory mapped I/O*. The partitioning approach is based on *dynamic programming* [Knud95, KnMa96] and estimates resource costs and execution time. The goal is to minimize the system execution time for a given hardware area constraint.

2.2.3 Single-Processor Architectures with many ASICs

These approaches support the design of a target architecture containing a set of ASICs (see figure 2.3) to speed-up the CPU. Examples of co-design systems supporting multiple ASICs are TOSCA and VULCAN.

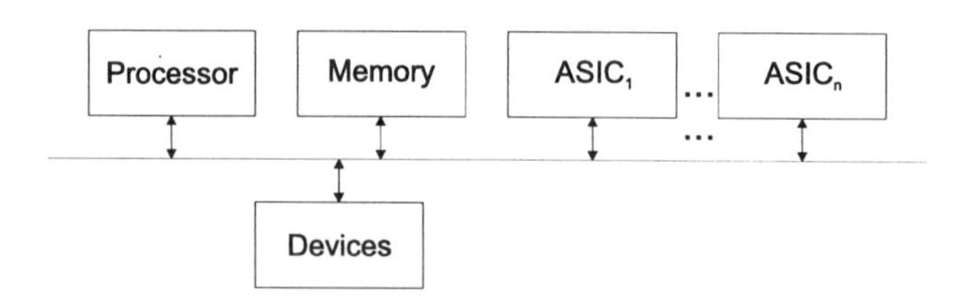

Figure 2.3. Single-processor architecture with many ASICs

[4]BSS: Braunschweig Synthesis System
[5]LYCOS : Lyngby Cosynthesis

Mickey. MICKEY [MGB93, MRB96a] is a knowledge-based hardware/software co-design tool for reactive systems supporting microprocessor-based hardware and software. Systems are specified using the visual specification language `SpeX` which is an augmented version of `StateCharts`. Then, MICKEY refines the specification into a control and data flow graph of *primitive functions* for which specific hardware and/or software implementations exist. These refinement steps use pre-compiled knowledge about the application domain, the transformation rules and the primitive functions. After the primitive functions have been computed, they are partitioned into a set of hardware and software implementations using an CLP[6] approach [MQB95]. Finally, the system is implemented using interface, software and hardware design steps. If a design goal cannot be fulfilled due to some knowledge hole, then the lower-level tool MINNIE [MRB96b] is invoked to find a solution. Summarizing, MICKEY and MINNIE represent a two layer co-design framework.

Tosca. TOSCA[7] [BFS95, BFS96] targets to reactive real-time systems which are control flow dominated. The target architecture is a single chip containing an off-the-shelf microprocessor core with its memory and a set of ASICs. Communication between hardware components is implemented through dedicated lines, communication between the microprocessor and an ASIC is based on memory mapped I/O. Systems can be specified using `C`, `VHDL` or `Occam`. A front-end compiles these system specifications into an object-oriented design database which is used by all tools. The internal representation is based on a customized *process algebra*. The partitioning approach is an iterative approach modifying the initial specification through the application of formal transformations provided by the process algebra. The partitioning can be iteratively executed controlled by the user or automated driven by some closeness metrics. The result of TOSCA is a `VHDL` description of all hardware including interface parts and the assembly code for the selected microprocessor.

Vulcan. VULCAN [GuMi96] is a hardware/software co-design tool similar to COSYMA focussing on co-synthesis. The target architecture contains one processor and one or more ASICs using all the same communication channel and memory. The input to VULCAN consists of a system specification described in the `HardwareC` language and design constraints including timing and resource constraints. Then, an internal flow graph representation is computed, used for performance estimation of hardware and software solutions. The automatic partitioning approach works iteratively, starting with a complete hardware solution. Then, parts of the system are migrated to the processor to reduce the amount of hardware area while still satisfying performance constraints. In con-

[6] CLP: Consistent Labeling Problem

[7] TOSCA : Tools for System Codesign Automation

trast to COSYMA, VULCAN is able to handle multiple processes as hardware and software running in parallel.

2.2.4 Multi-Processor Architectures

The target architectures in this category contain multiple processors, multiple ASICs and one or more communication channels. Figure 2.4 shows an example of such a *multi-processor architecture* with only one channel.

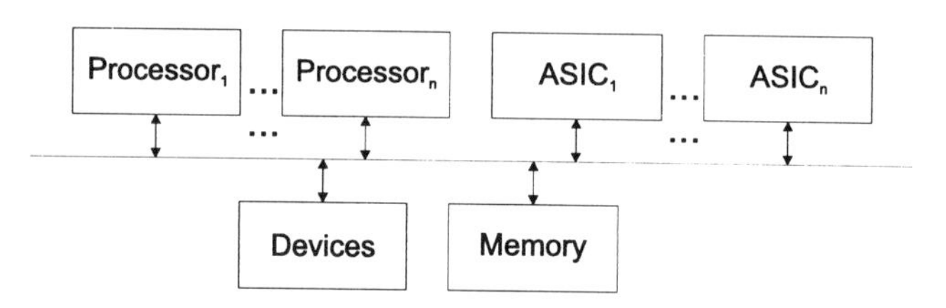

Figure 2.4. Multi-processor architecture

The approaches of this category which will be described in the following have different objectives: CHINOOK and COSMOS emphasize interface and communication synthesis, COWARE on heterogeneous modelling and implementation, POLIS supports co-verification and SPECSYN is a tool for design space exploration.

Chinook. CHINOOK [BCO95, COB95a] is a co-synthesis tool for embedded real-time systems. It is intended for control flow dominated designs constructed from off-the-shelf components. The main emphasis of CHINOOK is on interface and communication synthesis. In CHINOOK, the designer specifies the behavior of the system in `Verilog`, the timing/performance constraints and a set of structural components that will be used to implement the system. The components are selected from an architecture library containing devices and processor specifications. The designer determines the hardware/software partitioning manually and CHINOOK automates the most time-consuming and error-prone tasks like scheduling to meet timing constraints, communication and interface synthesis. This co-synthesis task is very difficult, because the protocols of the selected ports may initially be incompatible. One strength of CHINOOK is that incompatible protocols can be made compatible with additional hardware [OrBo97]. In addition, CHINOOK allows the designer to simulate the system specification and its implementation in the same simulation environment.

Cosmos. COSMOS [IsJe95] is intended to assist the designer in refining a system from system level to implementation by using powerful transformations. The key issue in COSMOS is communication synthesis. In a first step, the system is specified in `SDL`. The specification is then automatically translated into the intermediate format *Solar* [JeOb94]. All following design steps are working on the Solar format. The hardware/software partitioning is done manually by the designer supported by the partitioning tool PARTIF [IKJ94] allowing to apply a small set of transformation and decomposition rules. PARTIF decomposes, for example, a part of the system into a set of subsystems communicating through an abstract communication channel. In the following communication synthesis step, the designer selects an appropriate communication protocol for the abstract channel from a communication library. During the final architecture generation step, the resulting Solar representation is translated into a `C` or `VHDL` description, either for software or hardware parts. These design transformations have to be repeated until a complete and satisfactory implementation of the system is found. The result is a multiprocessor architecture implementing the specified behavior. The drawbacks of COSMOS are that evaluation tools and algorithms for partitioning and communication synthesis are missing. Therefore, optimization will be the key issue of future work.

CoWare. COWARE [RVBM96] is a system design environment supporting heterogeneous system specification and systematic refinement of the specification into a heterogeneous implementation. The target architecture contains multiple processor cores and ASICs. The heterogeneous system specification consists of communicating processes specified in high-level languages, such as `C`, `VHDL` or `DFL`. Functional behavior and communication are strictly separated to reuse specified processes. This separation is implemented by specifying communication between processes through ports which are connected by a channel. The communication semantics is based on *RPC*[8]. This heterogeneous system specification can then be simulated on a workstation. COWARE is not only a tool for heterogeneous system specification, but also for implementation. After simulation, the designer merges the processes into clusters and assigns them manually to processors of the target architecture. After this partitioning, the communication scheme is selected. Abstract communication channels are now refined by describing how this communication is executed. Then, all merged processes are compiled into a heterogeneous implementation. Finally, the protocols of all processors are made compatible, by producing additional hardware and software to interface their timing diagrams.

Polis. POLIS [CEG+96] is a co-design tool targeting towards real-time reactive systems which are control flow dominated. The target architecture supported by POLIS contains general-purpose processors combined with a few ASICs and other components, e.g. DSPs. POLIS is working on a uniform and formal internal representation called *co-design finite-state machine (CFSM)*. These CFSMs are generated from a system specification described in `Esterel`. A formal verification approach is integrated which translates the co-design

[8]RPC: Remote Procedure Call

FSMs into the FSM formalisms. Afterwards, verification systems prove the correctness for these FSMs. Hardware/software partitioning is done manually by the designer while POLIS assists the designer by providing estimation tools. In this step, each CFSM is assigned either to a hardware or software implementation which are generated automatically by POLIS. Interfaces between hardware and software are synthesized within POLIS. In addition, POLIS generates a small real-time operating system consisting of a scheduler and drivers for the I/O channels. Communication can be implemented using I/O ports available on the microcontroller or general memory mapped I/O.

SpecSyn. SPECSYN [GVN94, GaVa95, GVNG96] is a system-design environment supporting target architectures containing multiple processors, ASICs, memories and busses. It has been developed with the goal of providing extensive designer interaction. Systems are specified using the `SpecCharts` language which is a combination of `StateCharts` and `VHDL`. SpecCharts is based on the *program-state machine* (PSM) model. The specification is translated into an intermediate representation called *SLIF*, on which all SPECSYN tools are working. The SPECSYN approach to system design consists of three tasks: *allocation*, *partitioning*, *refinement*. During the allocation step, target architecture components are allocated. Processors and ASICs are allocated for computational parts of the specification, memories for variables which have to be stored and busses for communication channels. The partitioning step maps these objects now to the allocated target architecture components. Different partitioning algorithms are integrated in SPECSYN. Fast estimators produce a variety of system metrics for each partition, including hardware and software performance, bit-rates, hardware size, pins, software program and data sizes. After partitioning, the initial specification is refined automatically including memory allocation of shared variables and inserting interface protocols.

2.2.5 Other Co-Design Approaches

The above overview of co-design approaches supporting heterogeneous implementation is only a part of existing work in this area. The interested reader is referred to further approaches: ADEPT [KAJW93], CODES [BuVe92], CODE-SIGN [Esse96], MSCE [Calv93, Calv96], SIERA [SrBr91].

A complete list of up-to-date co-design approaches can be obtained via internet on

```
http://ls12-www.informatik.uni-dortmund.de/~niemann/codesign_links.html
```

2.3 Comparison of Co-Design Approaches

All described approaches can be compared by their specification and implementation power. The *specification power*, depicted in figure 2.5, classifies the

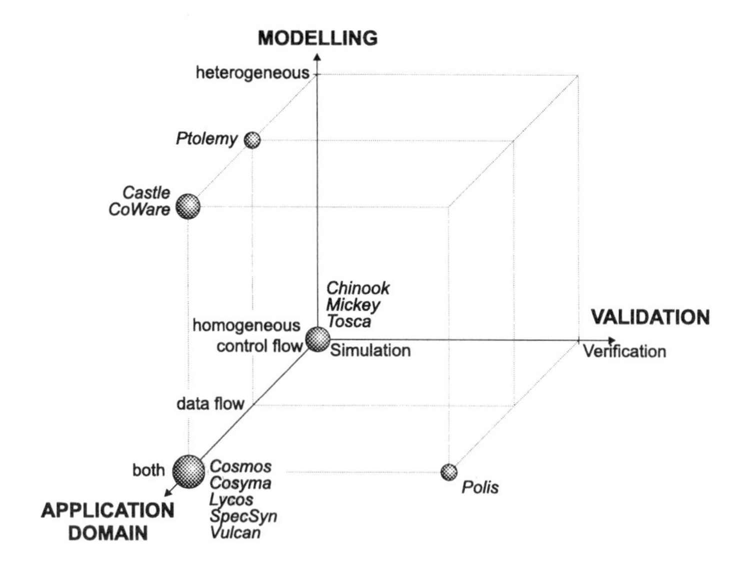

Figure 2.5. Categorization by specification power

different co-design tools concerning their modelling style (homogeneous/heterogeneous), their features for validation (simulation or verification) and their application domain (control flow, data flow or both). It can be summarized that all approaches offer a simulation possibility, but only POLIS supports co-verification. Most of the approaches use homogeneous system specification, only three of eleven tools support a heterogeneous one. A key result is that there is no tool supporting heterogeneous system specification and co-verification. This is not surprising, because these two goals are contrary. The usage of multiple specification languages for different parts of a system results in a loss of analytical power. In this case, co-verification becomes much harder.

The second criterion to compare the co-design tools is *implementation power* (see figure 2.6). Implementation power is classified by the complexity of the target architecture, the degree of automating the hardware/software partitioning and the support of implementing interfaces between processors and ASICs. The target architectures are divided into the following classes: single-processor-single-ASIC architectures with exclusive *(1,1,x)* or concurrent *(1,1,c)* execution, single-processor-multiple-ASIC architectures *(1,n)* and multiple-processor-multiple-ASIC architectures *(n,n)*. The support of implementing interfaces is divided into three classes (*low, medium, high*) corresponding to the degree of integrated techniques for interface and/or communication synthesis. As a key result it becomes clear that there is no tool which performs automatic partitioning and additionally offers a flexible support of interface implementation techniques. Most of the approaches emphasize one of both problems.

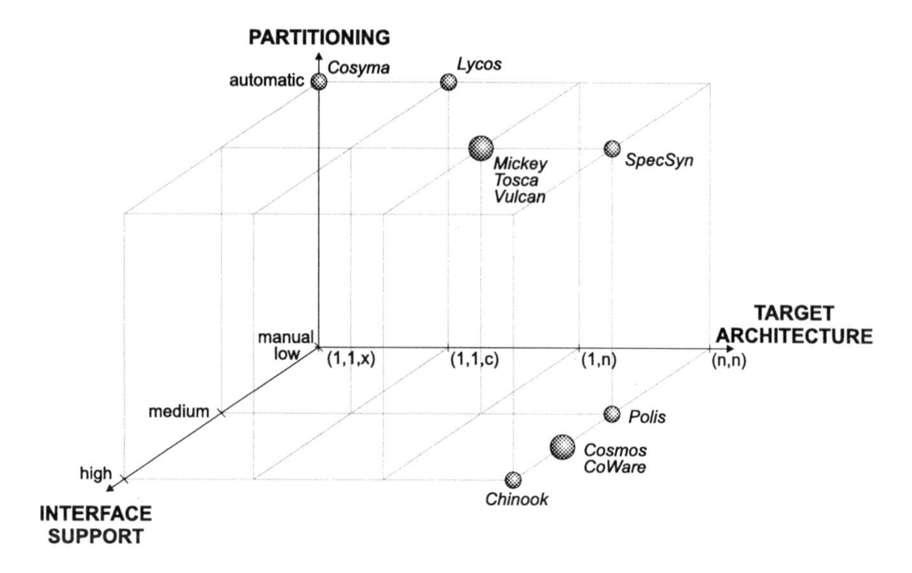

Figure 2.6. Categorization by implementation power

2.4 Overview of COOL

In this book the co-design tool COOL will be described which has been developed for data flow dominated systems. An overview of COOL is given in figure 2.7 COOL uses a homogeneous system modelling approach using a subset of VHDL for specification. A graphical user interface has been developed to specify these systems in a structural and hierarchical way. These systems are stored in a *system library*. In addition, the graphical user interface is used to define *target architectures* and *design constraints*. The target architectures are organized in a *target architecture library* too. The main objective of COOL is heterogeneous implementation. Several algorithms for hardware/software partitioning have been developed allowing the designer

- to compute optimal solutions (by using an integer programming approach)
- to compute high-quality solutions in acceptable computation time (by using a combination of integer programming and a heuristic)
- to perform design space exploration (by using a genetic algorithm approach)

In contrast to all other approaches using estimation algorithms during hardware/software partitioning, COOL uses the synthesis and compilation tools to compute the value for the cost metrics. The software parts are compiled using the compilers for the target processor and the ASICs are synthesized by using high-level synthesis. As a consequence, COOL works on high-precision estimates which are stored in *cost libraries*. The usage of a cost library supports reuse and therefore computation time is saved.

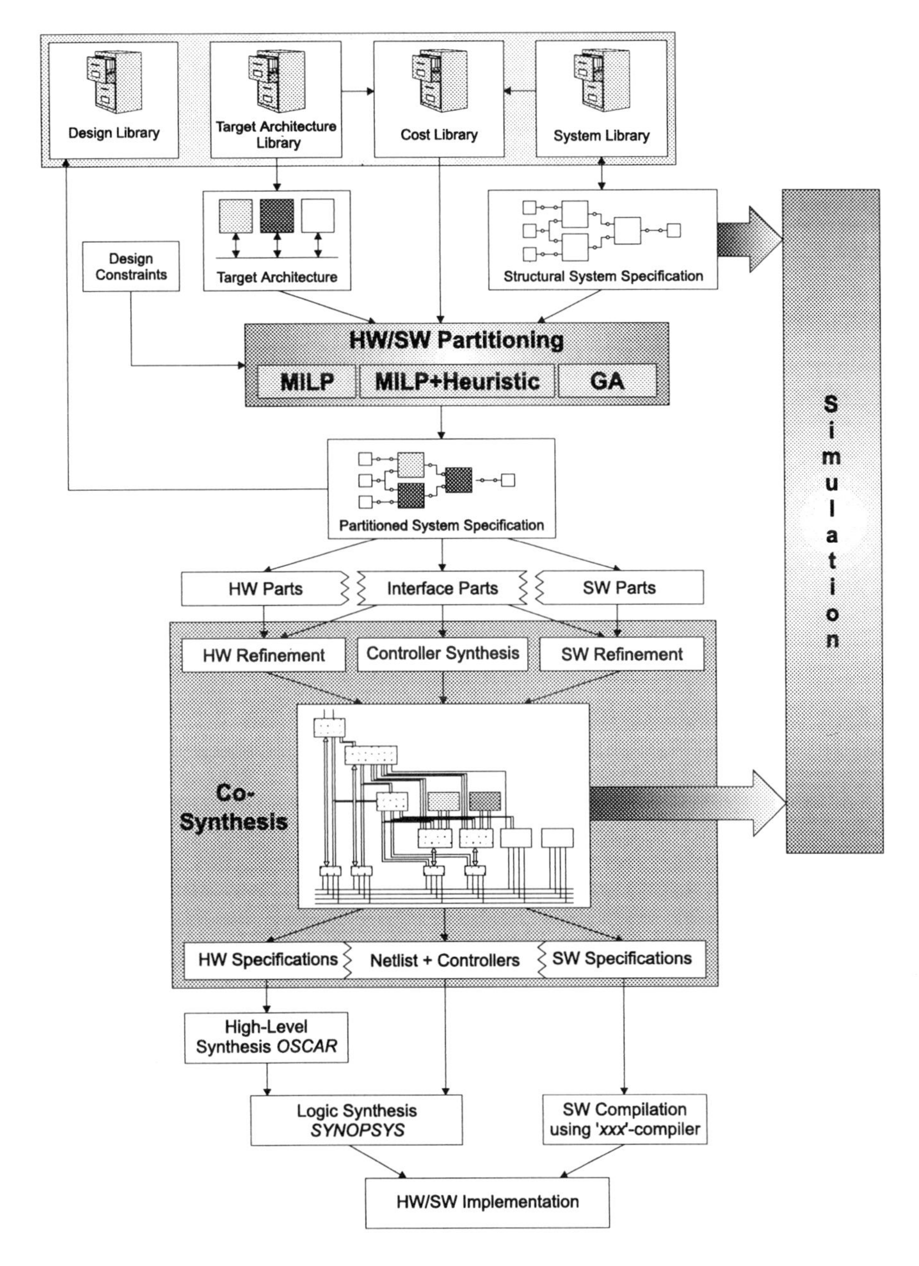

Figure 2.7. Overview of COOL

All partitioning algorithms integrated in COOL compute a mapping of the components of the system to processors and ASICs and allocate communication channels for required interfaces between hardware and software parts. In addition, a schedule is calculated defining the execution order of the system components on their resources.

The co-synthesis algorithm of COOL refines the initial specification by

- implementing the abstract communication channels by adding dedicated lines to the hardware components and additional software for processors to implement communication based on memory mapped I/O,
- adding a *system controller* implementing a *run-time scheduler* steering the complete system according to the computed schedule,
- adding an *I/O controller* to handle input and output values,
- adding a *bus arbiter* to prevent bus-conflicts,
- generating a complete netlist for wiring all components.

The result of COOL is a VHDL netlist which can directly be synthesized by the high-level synthesis tool OSCAR and the logic synthesis tool from SYNOPSYS. In addition, assembler code is generated for the processors which has to be down-loaded.

To validate the correct functionality of the system specification and its implementation after co-synthesis, the commercial VHDL simulator VANTAGE OPTIUM has been integrated in COOL.

2.5 Classification of COOL

In figure 2.8, COOL is categorized with respect to existing co-design approaches considering its specification power. As mentioned before, VHDL is used in COOL for system specification in a homogeneous manner. The use of VHDL allows validating the correct functionality of the specification by simulation but not by verification.

Concerning implementation power, COOL has to be categorized as depicted in figure 2.9. COOL supports the design process for multiple-processor-multiple-ASIC target architectures. The partitioning and co-synthesis phases work fully automatic and allow the implementation of hardware/software systems communicating via both message passing and shared memory mechanisms.

In the following chapters, the different co-design steps including system specification, hardware/software partitioning and co-synthesis will be described. Both the approaches published in the literature and the approach taken in COOL will be explained.

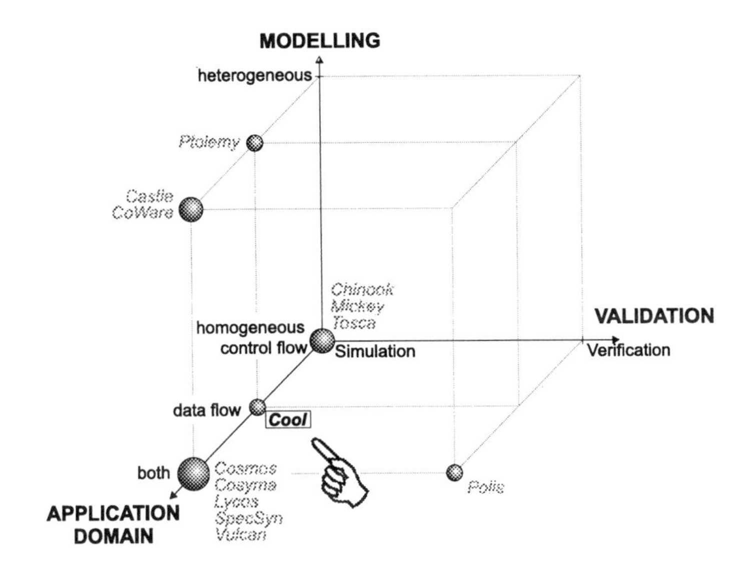

Figure 2.8. COOL categorized by specification power

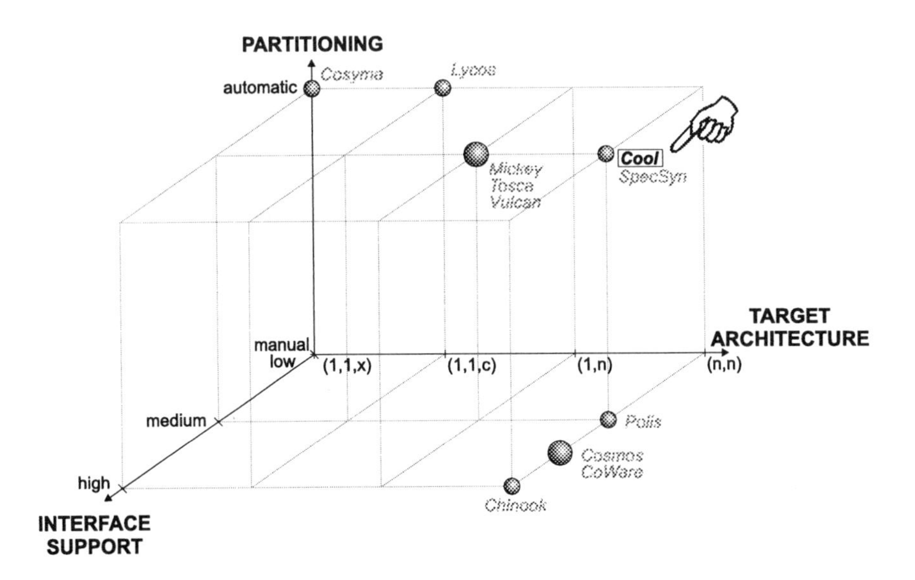

Figure 2.9. COOL categorized by implementation power

3 SPECIFICATION OF EMBEDDED SYSTEMS

At the beginning of the hardware/software co-design process, the designer has to specify the functionality of the system. This step is called *system specification.* In practice, two different approaches of specifying the functionality of an embedded system are adopted. The first approach is called *homogeneous modelling,* using one single language for specifying the system. The specification should be implementation-independent without favouring a hardware or software implementation. The typical co-design task using a homogeneous modelling approach is to analyze and allocate parts of the specification being implemented in hardware or software. In contrast, *heterogeneous modelling* uses specific languages for hardware parts (e.g. `VHDL`) and software parts (e.g. `C` or `C++`). This method of specifying systems allows a simple mapping to hardware and software, but the heterogeneity makes the analysis and validation of the system functionality much more difficult. In the following, system specification is referred to the meaning of homogeneous modelling.

The functionality of a system can be described with different methods. Each of these particular methods is called a model. A *model* has to be

1. *formal* to simplify the verification and analysis of the specified function. Furthermore, formal models do not contain ambiguity.
2. *complete* to describe all required information to design the system.
3. *comprehensible* and *easy to understand, to capture* and *to modify.*

4. *hierarchical* and *modular*, to handle the complexity of the systems.
5. *abstract* to describe the functionality independent of the implementation. Enough freedom should be left for algorithmic alternatives during the design process.

In the following, an overview of different models of computation will be given in section 3.1. Section 3.2 describes the requirements for specifying embedded systems and a set of existing specification languages will be compared in section 3.3. In section 3.4, the advantages and disadvantages of using VHDL for specifying systems will be discussed. At the end, the system specification method used in COOL will be introduced in section 3.5.

3.1 Models of Computation

Gajski [NVG92] classifies the following models of computation:

1. state-oriented models,
2. activity-oriented models,
3. structure-oriented models,
4. data-oriented models,
5. heterogeneous models.

These models will be described briefly in the following.

3.1.1 State-Oriented Models

State-oriented models describe the function of a system by a set of states and a set of transitions between them. These state-transitions react to external events. Therefore, the main application area is the domain of control flow dominated applications, e.g. reactive systems. Examples of state-oriented models are finite-state machines (FSM), Petri nets, and hierarchical concurrent FSMs (HCFSM).

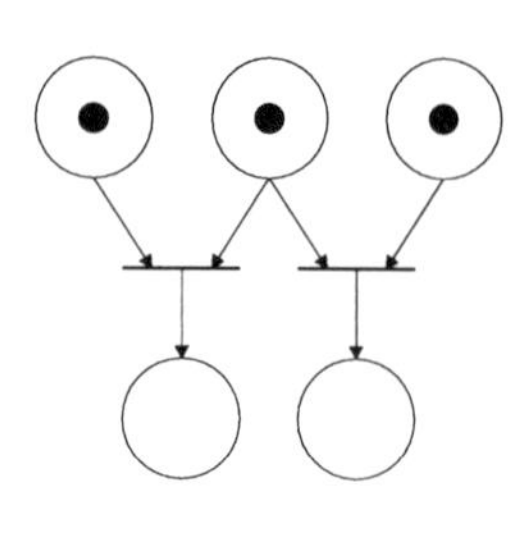

Figure 3.1. Petri net

Petri net [Reis85], as depicted in figure 3.1, is a graphical language for modelling complex systems. They have been developed by C.A. Petri [Petr62] in the early sixties. It was the first general theory to formulate discrete parallel systems. During the last decades a lot of different variants of Petri nets have been developed. In general, a Petri net consists of places, transitions and tokens. The tokens which are stored in places are consumed and produced whenever a transition fires. Therefore, Petri nets are well suited to model and analyze concurrent systems, e.g. for finding deadlocks.

In practice, Petri nets have proven to be useful for describing protocols typically used in networks. The drawback of Petri nets is that they become incomprehensible for complex systems.

Finite-state machines (FSM) [HoUl79] (figure 3.2) consist of a set of states, a set of transitions connecting states and a set of actions. One of the states is the initial state. There are two well-known types of finite-state machines: the *Mealy-automaton* is a transition-based FSM which associates actions with transitions. In contrast, the *Moore-automaton* is a state-based FSM associating actions with states. Moore-automata result in an increased number of states compared to Mealy-automata, because their output depends only on the states. However, the output of Mealy-automata depends not only on the states but also on the inputs. To reduce the number of states, these FSMs are sometimes extended by using variables and expressions. Finite-state machines are very well-suited for control flow dominated systems, because the system state is always defined explicitly. The main disadvantage of FSMs is the explosion of states and transitions when representing complex systems due to the lack of hierarchy and concurrency.

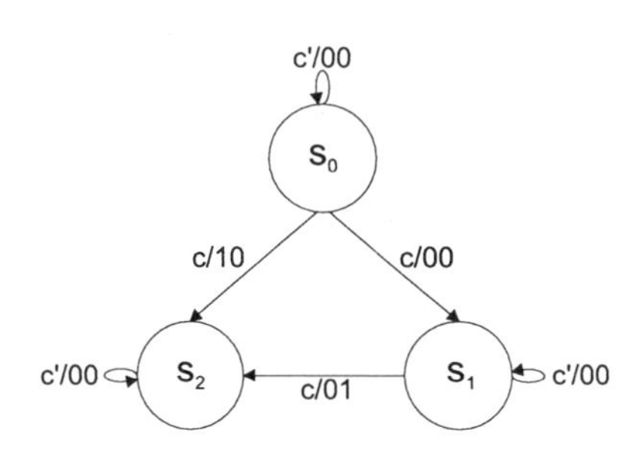

Figure 3.2. FSM

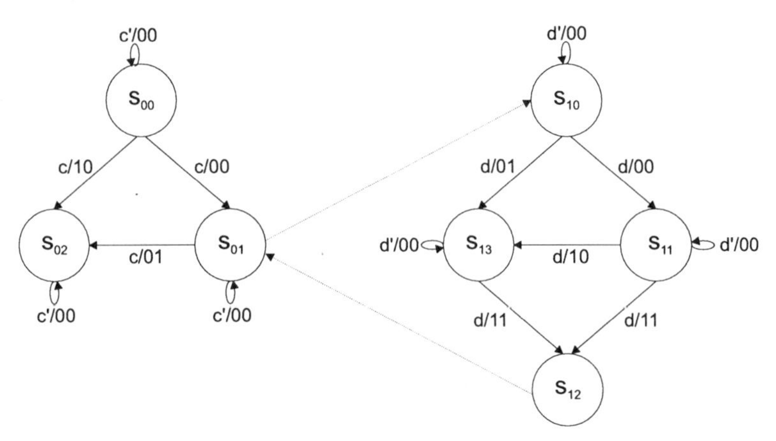

Figure 3.3. Hierarchical concurrent FSM (HCFSM)

Hierarchical concurrent FSMs (HCFSM), as depicted in figure 3.3, solve the main drawbacks of FSMs by decomposing states into a set of sub-states. These sub-states may be concurrent sub-states communicating via global variables. Therefore, HCFSMs support both hierarchy and concurrency and are well suited for representing complex systems. The disadvantage of HCFSMs is that control aspects can be easily modelled but not data and activities.

3.1.2 Activity-Oriented Models

Activity-oriented models specify the system functionality by a set of activities related by data or control dependencies. In contrast to state-oriented models, they have no internal state. Therefore, these models are very often applied to data flow dominated applications, as known, for example, in *digital signal processing.* The classical examples of activity-oriented models are data flow graphs (DFG) and control flow graphs (CFG).

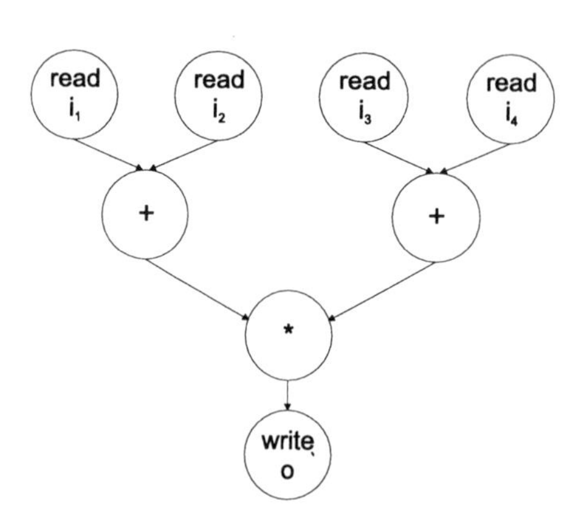

Figure 3.4. Data flow graph (DFG)

A *data flow graph* (figure 3.4) is a directed graph DFG(V,E) containing vertices and edges. Vertices correspond to actors representing inputs, outputs, storages and operations. Edges represent data dependencies between nodes. Operation nodes may also be complex functions, also being represented by a data flow graph. Therefore, DFGs support hierarchy and concurrency and they are well-suited for specifying complex *transformational systems.* The disadvantage of DFGs is that control cannot be modelled with them.

A *control flow graph* (figure 3.5) is a directed graph CFG(V,E) where the vertices correspond to basic blocks or decision nodes and the edges represent the flow-of-control between the nodes. A *basic block* is a sequence of statements without branches. A *decision node* is used to steer the flow-of-control into alternative directions. Control flow dominated systems can be modelled with them, but no data dependencies.

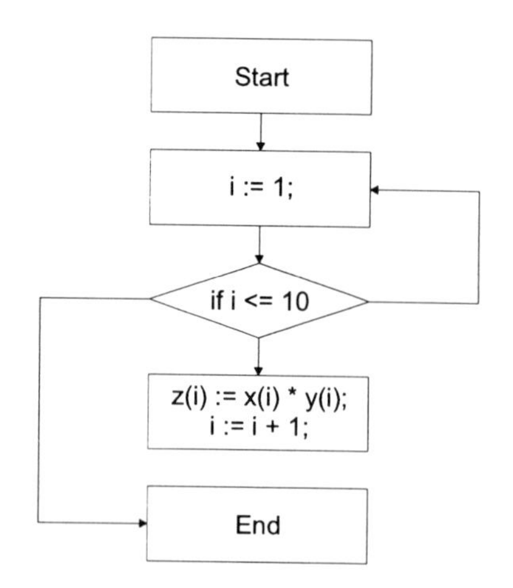

Figure 3.5. Control flow graph (CFG)

3.1.3 Structure-Oriented Models

Structure-oriented models are often used in later design phases. In contrast to state-oriented and activity-oriented models which describe the system functionality, the structure-oriented models describe the physical modules and intercon-

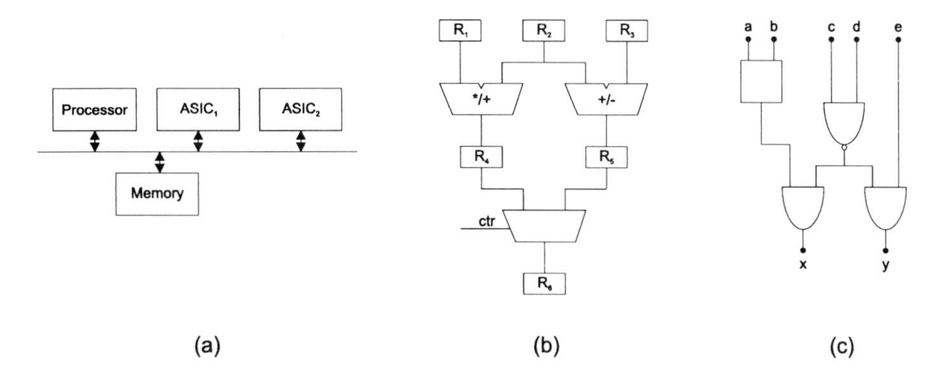

Figure 3.6. Block diagram (a), RT netlist (b) and gate netlist (c)

nections of the system. Typical examples are *block diagrams* (figure 3.6a), *RT*[1] *netlists* (figure 3.6b) and *gate netlists* (figure 3.6c). Very often, these structure-oriented models are hierarchical, so that a system-block is also a structural model containing a set of sub-systems and interconnections.

3.1.4 Data-Oriented Models

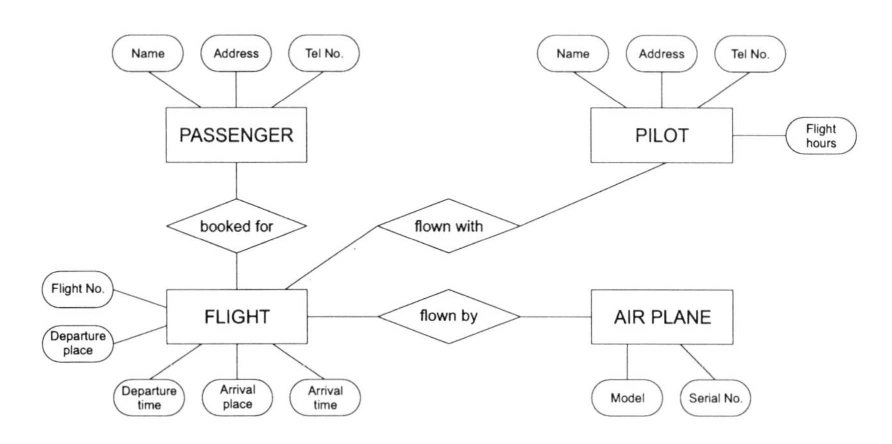

Figure 3.7. Entity relationship model

Data-oriented models offer a totally different view of describing systems. In contrast to all other models described before specifying the activities of a system, data-oriented models define the relations between data which are consumed by the system. Data-oriented models are very useful for *informational systems*. The typical example of data-oriented models is an *entity-relationship diagram*

[1]RT : Register Transfer

[Chen77], as depicted in figure 3.7. It defines the consumed data of a system with entities and relations. Each entity (e.g. 'passenger') contains a set of attributes (e.g. 'name', 'address', 'telephone number') storing the interesting information of the entity. A relation (e.g. 'booked for') reflects some fact ('flight x is booked for passenger y') relevant to its entities ('passenger', 'flight'). In summary, entity relationship diagrams provide a good view of data consumed in the system and their relations to each other. The main disadvantage is that functional behavior cannot be expressed by them.

3.1.5 Heterogeneous Models

Heterogeneous models combine special characteristics of the four different models presented before.

Control/Data flow graphs (figure 3.8), for example, combine the advantages of data flow graphs and control flow graphs in one model. The nodes of the control flow graph are now represented by data flow graphs. Therefore, a CDFG[2] represents not only the control flow of a system, but also the internal data dependencies. CDFGs are very often used as an intermediate format, e.g. in high-level synthesis.

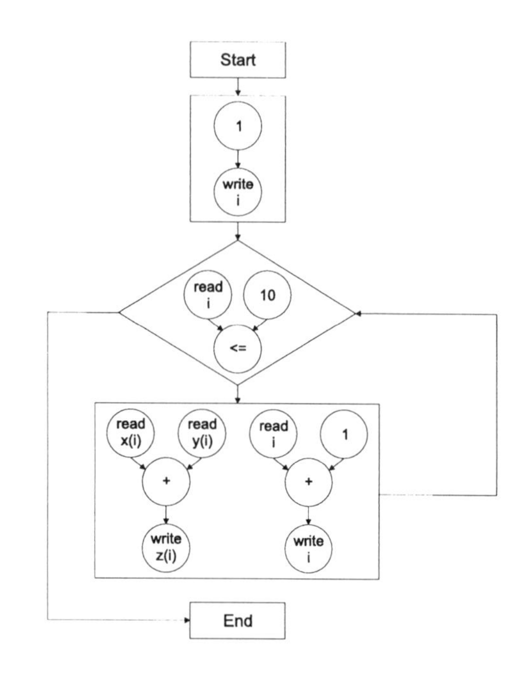

Figure 3.8. Control/data flow graph (CDFG)

Programming languages combine aspects of data-oriented models (e.g. definition of data-types and declaration of variables, etc.) and activity-oriented models (e.g. sequential statements, branches, loops, functions, procedures, etc.). There are two major classes of programming languages: imperative and declarative languages. In programs implemented with *imperative languages* (e.g. `ANSI-C` [KeRi78], `Pascal` [Wirt71]) the statements are executed in the same order as defined in the specification. In contrast, *declarative languages* (e.g. `Prolog` [StSh86]) specify no explicit sequence of executing program statements. The sequence of computation in declarative languages is driven by a

[2]CDFG : Control Data Flow Graph

set of functions or logic rules. The general disadvantage of most programming languages is that they have no special language constructs to model a system's state.

Object-oriented models [Booc91] are a combination of data- and activity-oriented models. Objects are the basic elements of these models consisting of a set of data and a set of operations transforming this data. These objects execute their task independently, representing a natural concurrency. New aspects of object-oriented models are *data abstraction* and *information hiding.*

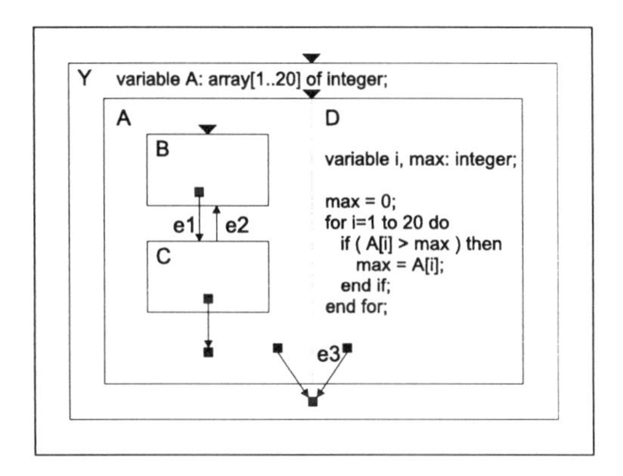

Figure 3.9. Program state machine (PSM)

A *program-state machine* (PSM) [VNG91] combines the principles of hierarchical finite-state machines with programming language constructs. An example of a program-state machine is given in figure 3.9. A program-state machine consists of a hierarchy of program states executing a special mode of computation. A program state may be a composite program state (consisting of concurrent or sequential sub-states) or a leaf program state (representing a leaf in the hierarchy tree). These leaf program states are described using programming language constructs. PSMs overcome the limitations of programming languages and HCFSM models. The main disadvantage of program-state machines is that this model is not based on formal semantics. Therefore, formal verification methods cannot be applied.

3.2 Requirements for Embedded System Specification

The presented models of computation focus on different aspects of specifying a system. The optimal system specification language is able to specify the complete functionality of the system while minimizing the design effort. In this section, the requirements for specifying embedded systems will be presented. The nature of embedded systems leads to a set of required characteristics which have to be supported by specification languages. These characteristics consider the expressive power of the language, analytic aspects and commercial aspects.

The *expressive power* of languages used for specifying embedded systems include the following aspects, most of them summarized in [GVNG94]:

1. *Hierarchy* is a very important feature of system specification languages to handle complex systems. It simplifies the specification process by enabling *top-down* or *bottom-up* specification. This hierarchy may be a structural or a behavioral hierarchy. *Structural hierarchy* enables the designer to compose a system of interconnected components which are themselves composed of sub-components. *Behavioral hierarchy* decomposes a behavior into a set of sub-behaviors, e.g. procedures or functions.

2. *Programming constructs* are necessary to simplify the specification. Complex data types (records, arrays, lists, etc.) are useful to describe the consumed data in a comfortable way. Constructs like functions, procedures, loops, branches (if, case) and assignments simplify the description of the sequential parts of the system behavior.

3. *State-Transitions* are required to model the internal state of an embedded system which is very important for control flow dominated or reactive systems. State-transitions could be modelled by programming constructs (e.g. using a case-construct), but they make the system specification much more difficult to read.

4. *Behavioral completion* is required to determine when a behavior has executed all computations. This may be useful to start another behavior in sequence of this finished behavior.

5. *Concurrency* has to be expressible, because very often parts of embedded systems are working concurrently. These parts may work in parallel at different levels of abstraction, like jobs or tasks, for example. At the process level, threads of the processes working independent of each other should also work in parallel.

6. *Communication* and *synchronization* are required when the concurrently working parts exchange data. Communication mechanisms like the *shared memory* (using a shared medium to exchange data initiated by the sender) or the *message passing* model (using abstract communication channels with send/receive primitives) are necessary. Both shared memory and message passing models can be transformed into the other model. To synchronize this data exchange additional handshake protocols and status signals are required. If a process P1 is waiting for a response of another process P2, both can be synchronized by using events or data. In the first case, P2 sets an event and activates P1. In the other case, P2 writes a special value into a variable which is read by P1.

7. *Exceptions* (e.g. interrupts or resets) may occur in embedded systems, because embedded systems often implement reactive systems communicating with the environment. The current system state has to be terminated and a transition into a new state is required. In cases of interrupts the original

state has to be continued afterwards. Specification languages should be able to model these exceptions and the reactions activated by them.

8. *Timing aspects* are very important to specify an embedded system, because very often these system have some strict requirements concerning the timing, e.g. the response time of an anti-lock brake system (ABS) in automobiles. Therefore, functional timing and timing constraints have to be specified. *Functional timing* represents the consumed time for executing a behavior. In contrast, *timing constraints* specify a range of elapsed time for executing a special behavior. Timing constraints include, for example, minimum/maximum timing or rating constraints.

9. *Non-determinism* is useful if the designer does not want to specify some special aspects of the system leaving details to the implementation step. If, for example, two external events occur simultaneously, it might be better if the implementation step determines the execution order of the initiated behaviors instead of defining a fixed schedule in the specification.

10. *Environment characteristics* are non-functional aspects of the specification, but important for embedded systems communicating intensively with the environment. This environment should be specified by a set of properties, related to operational conditions such as timing specifications of incoming data (frequency, timing, waveform type).

In contrast to the characteristics considering the expressive power of specification languages, analytical aspects are important, to analyze the behavior of the system at specification time, before designing it. *Analytical aspects* include the following:

1. *Verifiability* and *formal analysis* are key issues of analytic aspects. Formal specification languages (e.g. LOTOS) based on a formal interpretation model, enable the designer to verify the specification, e.g. to detect *dead-locks*.

2. *Model executability* is another possibility beside formal analysis of analytical aspects. Model executability is gained by executable specifications which are of great importance. They offer the possibility to experiment with different alternatives of the system specification, e.g. by simulating the specification with a set of stimuli. One goal is to validate and analyze the correct function of the system. Another goal may be the detection of computational-intensive parts in the specification which might become the bottleneck of the final design.

Finally, commercial aspects are of great importance, because using a certain specification language has a great impact on the time-to-market of the products. *Commercial aspects* include the following:

1. *Availability* of tools supporting these languages, e.g. simulators, debuggers, compilers (for different processors) and synthesis tools (for different hardware

libraries) is very often the most important aspect in industry, because short time-to-market is one key goal.

2. Other commercial aspects consider the *standardization* of a specification language. Standardized languages are well-defined, supported by tools and in most cases (e.g. VHDL) improved continuously.

3. *Simplicity* and *clarity* may also be an important point to choose a specification language, because a clarified specification may serve as a good documentation of the system.

3.3 Survey of System Specification Languages

The overall goal of a specification language is to describe the desired functionality of a system non-ambiguously. A good system specification language has to fulfill all required characteristics described before, but until now, there is no specification language being the best for all possible applications. Research in the area of system specification languages is going on intensively. A detailed analysis of different specification languages applied to embedded systems is beyond the scope of this book. Therefore, only a brief overview and comparison of different specification languages in this area will be given. The interested reader is referred to [Berg95, Calv93, NVG92]. In recent years, a lot of different specification languages have been developed for different application domains:

1. *Formal description techniques*:

   - LOTOS[3] [EVD89] is a language for specifying concurrent and distributed systems. It has been used extensively in the telecommunication industry for describing OSI[4] protocols. LOTOS is based on *process algebra* and has been standardized [ISO89] by the ISO[5] in 1989.

   - Another, probably the best-known specification language from the telecommunication area is SDL[6] [BHS91, DFM+97]. It has been applied to describe distributed real-time systems at a very high level of abstraction. It is based on *extended finite-state machines*. In 1988 and 1992 it has been standardized by the ITU[7] [ITU88, ITU92].

   - A third formal description language in the area of telecommunication is Estelle also based on extended FSMs. Estelle (standardized [ISO87] by ISO) is very similar to SDL, but in the last years SDL gained more importance.

[3] LOTOS : Language of Temporal Ordering Specifications
[4] OSI : Open Systems International
[5] ISO : International Standard Organization
[6] SDL : Specification Description Language
[7] ITU : International Telecommunication Union

These three languages are described and compared in [Hogr89].

2. *Real Time System Languages*:

   - **Esterel** [Berr92, BeGo92] is a *synchronous programming language* developed for specifying reactive real-time systems. Synchronous programming languages are based on *perfect synchrony hypothesis* meaning that statements take zero execution time. As a consequence, timing constraints cannot be modelled in **Esterel**. **Esterel** is a very simple FSM model based language including constructs for hierarchy, pre-emption and concurrency.

   - **StateCharts** [Hare87] is a graphical specification language extending classical finite-state machines by hierarchy, timing specification, concurrency and communication/synchronization mechanisms. It has been developed for specifying reactive systems. The main drawback of **StateCharts** is that programming constructs are missing.

   - *High-Level Time Petri Nets* [Esse96] are a combination of high-level Petri nets [Jens90] and time Petri nets [MeFa76]. In *high-level Petri nets* the type of tokens is extended to complex data types, e.g. boolean, integer, real, arrays, records and others. Also, the firing rule is extended. Transitions may fire depending on the result of a Boolean function on the input token data. The output token data values are computed by applying net inscriptions to the input token data values. Thus, data flow and control flow can be described with high-level Petri nets. *Time Petri nets* are able to model most temporal constraints, because a finite enabling duration is associated with each transition. This enabling duration is specified by a tuple (t_1, t_2) denoting the earliest (t_1) and the latest firing time (t_2) after the transition is enabled.

3. *Hardware description languages*:

   - **VHDL**[8] is a hardware description language supporting the hardware development process. It has been standardized twice until now [IEEE87, IEEE93], but the development of extensions to **VHDL** is still going on. **VHDL** will be described in more detail in section 3.4.

   - Another hardware description language is **Verilog** HDL[9] [ThMo91]. It has been developed to design and document electronic systems at various levels of abstraction. In 1996, it has been standardized [IEEE96]. In contrast to **VHDL**, exceptions can be modelled by using a disable-statement. State-transitions are not supported.

---

[8] **VHDL** : VHSIC (Very High Speed Integrated Circuit) Hardware Description Language

[9] HDL : Hardware Description Language

- HardwareC [KuMi88] was specially designed for hardware synthesis. It extends the ANSI-C [KeRi78] programming language by additional constructs and well-defined semantics including structural hierarchy, concurrency, communication and synchronization. Timing constraints can be defined between pairs of statements using labels. A special feature for hardware synthesis is the possibility of specifying resource constraints.

- SpecCharts [NVG92] has been developed for the specification of embedded systems. It is based on *program-state machines* combining the advantages of hierarchical finite-state machines with those of the programming language VHDL.

4. *Programming Languages*:

- Similar to HardwareC, another extension of the ANSI-C language is the language C^x [EHB93]. C^x has been developed for the specification and simulation of embedded systems. It extends C by additional timing constraints and communication between tasks.

- Recently, other approaches have been published using C-based languages for system specification. SpecC [ZDG97] has been developed to improve traditional HDLs such as VHDL. It models a system in form of a hierarchical network of behaviors and channels. Concurrency, state-transitions, structural and behavioral hierarchy, exception handling, timing, communication and synchronization are supported.

Another C-based approach has been presented in [GuLi97]. The advantage of these C-based approaches is always that a large amount of existing code can be used.

5. *Parallel programming languages*:

- The CSP[10] [Hoar78] language is based on the *process algebra* with the same name. It has been developed to specify communication and synchronization between processes executed on multi-processor machines. To that time, traditional programming languages were limited to single-processor architectures. CSP allows the specification of a program as a set of concurrent processes communicating to each other.

- Occam [Inmo84] is a parallel programming language which has been derived from ideas of CSP. It is known as the language of *transputers*, but Occam is also often used to describe VLSI systems at several levels of abstraction. An Occam description is close to a CSP specification and therefore the correct functionality can be verified by a proof.

[10]CSP : Communicating Sequential Processes

6. *Data flow Languages*:
   `Silage` [Hilf85] is a data flow language which has been developed for *digital signal processing* (DSP) systems. `Silage` is an *applicative language* where computations for incoming signals are defined to compute output signals. In contrast to imperative languages where programs are represented as a sequence of assignments, a `Silage` program is a set of signal definitions. Consequently, the relative ordering of statements in the source code has no effect. `Silage` is a very restricted language without supporting standard programming language constructs.

In figure 3.10, an overview is given for some of the languages presented before and aspects for comparing their ability to specify embedded systems.

No.	Language	Structural Hierarchy	Behavioral Hierarchy	Programming Constructs	State-Transitions	Behavioral Completion	Concurrency	Synchr./Communication	Exceptions	Timing Aspects	Non-determinism	Formal Analysis	Model Executability	Availability of Tools	Standardization
1	Lotos	+	+	+	+	+	+	+	+	-	+	+	-	o	+
2	SDL	+	o	-	+	+	+	+	-	o	+	+	+	o	+
3	Esterel	+	+	+	-	+	+	+	+	-	-	+	+	o	-
4	StateCharts	-	+	-	+	-	+	+	+	o	-	-	+	o	-
5	High-Level Petri Net	+	-	-	+	-	+	+	+	+	-	+	+	o	-
6	VHDL	+	o	+	-	+	+	+	-	+	-	-	+	+	+
7	Verilog	+	+	+	-	+	+	+	+	+	-	-	+	+	+
8	SpecCharts	-	+	+	+	+	+	+	+	+	-	-	+	o	-
9	HardwareC	+	o	+	-	+	+	+	-	o	-	-	+	o	-
10	CSP / Occam	-	+	+	-	+	+	+	-	-	+	+	+	o	-
11	Silage	-	-	-	-	-	+	-	-	o	-	-	o	o	-

(+) fully supported (o) partially supported (-) not supported

Figure 3.10. Specification languages for embedded systems

As a conclusion, it becomes clear that until now, there is no single specification language supporting the whole range of applications. The first decision is whether the system should be specified in a homogeneous or a heterogeneous way. Then, the selection of the specification language(s) has to be selectively targeted, considering the application which should be specified, the expressive power, additional analytical and commercial aspects of the specification language.

3.4 VHDL for System Specification

VHDL is a hardware description language which has been developed in the early eighties for fulfilling a number of needs in the design process of digital electronic systems. In 1987, VHDL has been standardized by the IEEE [IEEE87]. In the last years, VHDL has gained great acceptance in industry and the language is continuously improved. Therefore, a re-standardization has been published in 1993 by the IEEE [IEEE93].

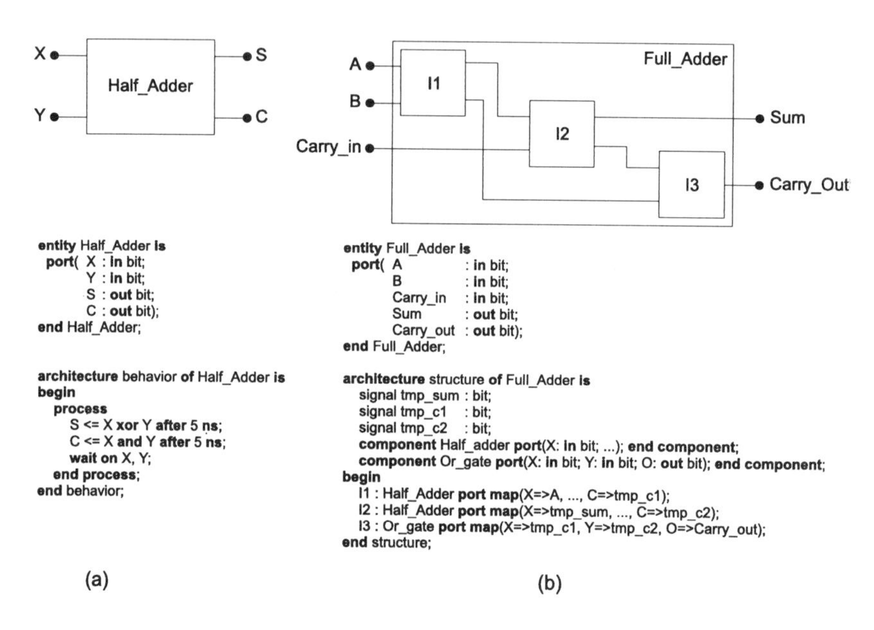

Figure 3.11. Behavioral (a) and structural (b) description in VHDL

A VHDL specification, called *design entity,* is defined by two parts: an entity declaration and an architecture body. The *entity declaration* is used to specify the interface of the design entity. The *architecture body* specifies the functionality of the design entity by a behavioral or structural description or by a mixture of both. A *behavioral description* (see figure 3.11a) contains a set of concurrently working processes executing statements sequentially. The statements are constructs of standard programming languages, e.g. assignments, if-statements, case-statements or loops. A *structural description* (see figure 3.11b) specifies the function of the entity by an interconnected netlist of sub-entities. Further information about VHDL can be found in [Ashe89, Coel89, Nava93]. Using VHDL for specifying systems has many advantages:

1. VHDL is a standardized language which is continuously improved.

2. VHDL is able to describe systems at different levels of abstraction, ranging from system level to gate level.

3. Several VHDL-based tools are widely available on different platforms ranging from PCs to workstations. In the last decades, commercial tools using VHDL have been developed for simulation (support of the VHDL IEEE 1076 standard) and hardware synthesis (support of a synthesizable subset of VHDL to generate a netlist).

4. VHDL specifications represent a good documentation, because they are executable.

The following features of VHDL concerning the expressiveness of system specification languages can be summarized:

1. VHDL supports *behavioral* and *structural hierarchy* allowing the designer to develop the system in a *bottom-up* or *top-down* methodology.

2. *Behavioral completion* and complex data structures (arrays, records, etc.) can be defined.

3. VHDL is an *imperative programming language* allowing the designer to use standard *programming constructs.*

4. *Functional timing* can be specified with VHDL, but no *timing constraints.*

5. *Synchronization* and *communication* can be modelled via signals.

6. *Concurrency* can be expressed by specifying processes working in parallel.

The main disadvantages of using VHDL are that no constructs exist for specifying a system's state or an incoming exception. States can be modelled indirectly by using a `case-statement` and a `state` variable, but there is no special language construct for states in VHDL. Nevertheless, several research groups in hardware/software co-design [Ecke93, EPD94, Madi96, NiMa97] use VHDL for specifying, because of its advantages summarized before.

3.5 System Specification in COOL

In COOL, systems are specified in a hierarchical manner using an integrated graphical user interface. The specification is based on the language VHDL. *System components* are specified using either behavioral or structural VHDL. The behavioral descriptions are specified textually. In contrast, the structural specifications are defined by wiring *instances* of pre-defined system components using the graphical user interface. A *system library* has been integrated to store specified systems in *library classes* allowing the designer to define the system either *top-down* or *bottom-up*. In figure 3.12, the specification technique is illustrated.

Example 1:

The specified system consists of two instances of component $fct1$, described behaviorally, and one instance of the structural component $fct2$. Component $fct2$ itself, consists of three instances of components $fct21$ and $fct22$. All these components are

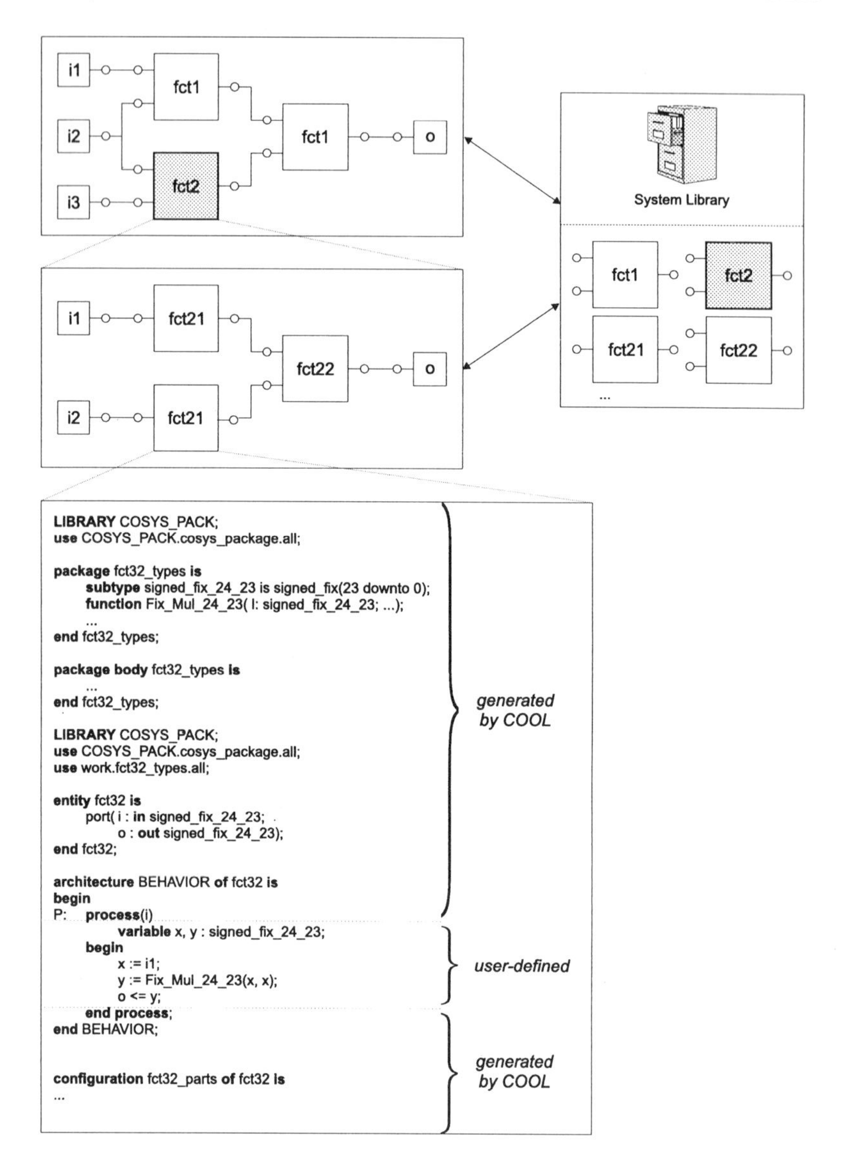

Figure 3.12. System specification using COOL

stored in the system library. To specify a system behaviorally, a designer first has to define the incoming and outgoing signals using COOL. Then, COOL generates a template including the required libraries and types for specifying the component. After this, the designer is able to define the behavior of the component by implementing the empty process statement. In the depicted case, $fct32$ returns the output o representing the square value of the input signal i.

Semantics. Semantics of a system specification is *data flow* oriented. The components communicate via signals, defined graphically by wires. These signals represent abstract communication channels. To be able to estimate the communication effort exactly, a component reads data using the incoming signals at the beginning of the process. Then, the component computes some results for the inputs. Finally, the results are written using the output signals. However, during computation no additional data can be read in, because otherwise the communication effort cannot be estimated exactly.

Granularity. At the lowest level of hierarchy, the system consists of wired behavioral `VHDL` specifications. In later design steps, each of these behavioral components is mapped to a processing unit of the target architecture. Therefore, the behavioral description represents the smallest granularity for partitioning which cannot be partitioned onto different processing units.

Restrictions. The behavioral specifications support only a subset of the complete `VHDL` language. The reason for this is that many constructs in `VHDL` cannot be synthesized in later design steps. The following restrictions have been applied for behavioral `VHDL` specifications in COOL:

- The supported data types are: `bit`, `integer`, `fixed-point`.
- Only one-dimensional arrays may be defined for these data types. To estimate the costs for variables and operations of these types exactly, the number of bits has to be specified precisely. An example is given in figure 3.12 where `signed_fix_24_23` represents a signed fixed-point number consisting of 24 bits and a fraction of 23 bits.
- Unbounded loops are not allowed, because the worst-case execution time cannot be estimated for them. Therefore, only `for`-loops with constant bounds are allowed.
- The timing construct `after` is not allowed, because it cannot be synthesized.

Some of these restrictions could be eliminated with some effort, but the main objective of this book is to find good algorithms for hardware/software partitioning and co-synthesis.

On the other hand, COOL tries to support the design process of data flow dominated systems having hard timing constraints, like audio- and video-applications. To simplify the specification of digital signal processing algorithms, special data types and functions are necessary, e.g. operations for `fixed-point` arithmetic. Therefore, a `VHDL` package (`cosys_pack.vhdl`) has been implemented for simulation, including all operations which might be used by the designer when specifying systems.

Simulation. A great advantage of COOL is the integration of the commercial VHDL simulator VANTAGE OPTIUM [Vant94] as illustrated in figure 3.13.

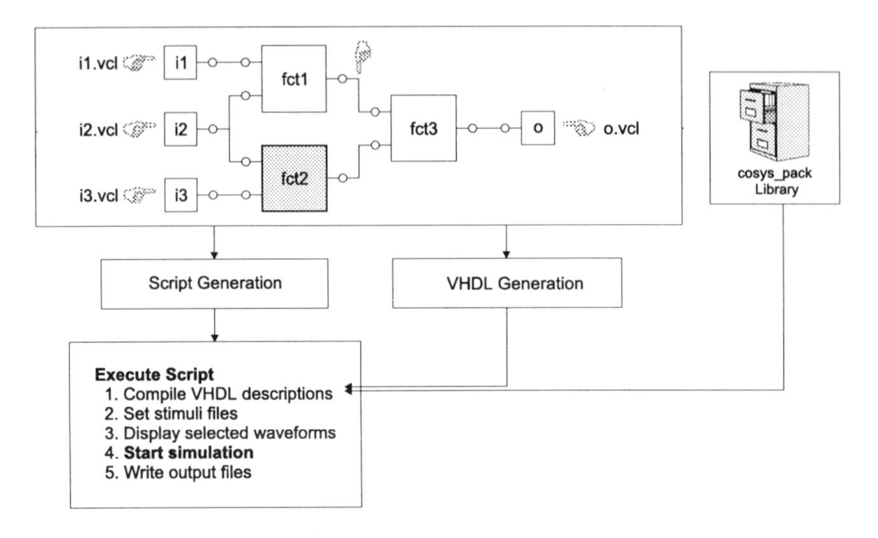

Figure 3.13. System simulation in COOL

Using the graphical user interface, the designer defines the *stimuli files* for the input signals and selects the signals for which waveforms should be depicted during simulation. Systems described behaviorally can be simulated directly. For structural systems a simulatable VHDL description is generated. In addition, a script is generated which is executed afterwards. This script steers the simulator, by compiling all necessary files including the VHDL package (`cosys_pack.vhdl`), loading the stimuli files for simulation, preparing and starting the simulation tools, etc. After the simulation has finished, the user can check the functional correctness of the specified system.

This interface between COOL and VANTAGE OPTIUM has been proven to be very helpful in practice [PG293b], because it automates time-consuming manual work. The integration of VANTAGE OPTIUM is not only used for simulating the system specification, but also for simulating the results of the design process, described in chapter 5. This allows the comparison between the simulation results of the specification and the results of the refined system specification after hardware/software partitioning.

Unified Design Environment. Another advantage of COOL is that all additional design information, including the definition of the *target architecture* and *design constraints* can be specified. Therefore, COOL represents a unified *design environment* for implementing hardware/software systems starting from a homogeneous, implementation-independent system-level specification. In particular, the specification of design constraints is very important, because the experience of the designer should be used during the design process. In COOL, a variety of different *constraints* are supported, including

- *mapping constraints*,
- *binding constraints*,
- *minimum* or *maximum timing constraints* (relative or absolute), and
- *resource constraints* (chip area, memory usage).

In figure 3.14, some of these user-defined constraints are depicted.

Example 2:

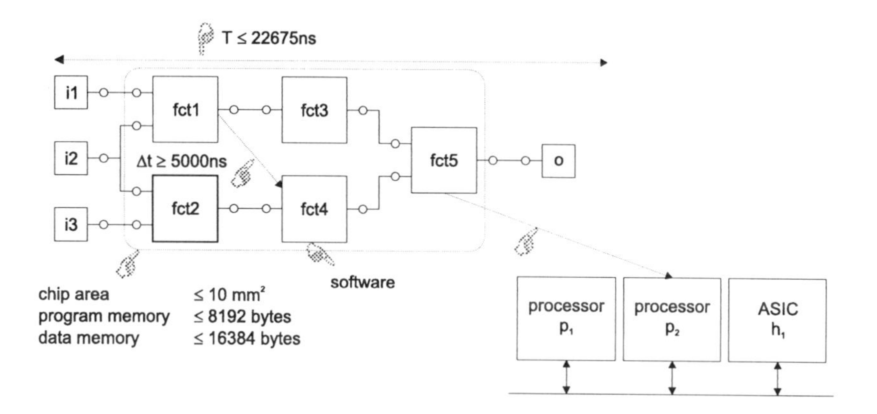

Figure 3.14. Design constraints specified with COOL

A mapping constraint has been defined for *fct4*. It should be implemented in software in general, either on processor p_1 or p_2. *fct5* has to be implemented on processor p_2 representing a binding constraint. Two different timing constraints have been specified. First, the complete system execution time T has to be smaller then 22675 ns (real-time condition for audio systems using a sample frequency of 44.1 kHz). Second, a relative timing constraint determines that *fct4* may not start earlier than 5000 ns after *fct1* has finished. Finally, resource constraints (chip area $\leq 10\text{mm}^2, \ldots$) have been defined for all components.

4 HARDWARE/SOFTWARE PARTITIONING

Embedded systems typically consist of application specific hardware parts and programmable parts, e.g. processors like DSPs, core processors or ASIPs. In comparison to the hardware parts, the software parts can be developed and modified much easier. Thus, software is less expensive in terms of costs and development time. Hardware, however, provides better performance. For this reason, a system designer's goal is to design a system fulfilling all performance constraints and using a minimum amount of hardware. The co-design phase, during which the system specification is partitioned onto the hardware and programmable parts of the target architecture, is called *hardware/software partitioning*. This phase represents one key issue during the design process of heterogeneous systems. In some co-design approaches, partitioning is done manually, but automatic hardware/software partitioning is of large interest, because the problem itself is a very complex optimization problem.

In the research area of hardware/software co-design a variety of different hardware/software partitioning approaches can be found. These approaches can be distinguished by the following aspects:

1. the complexity of the supported *partitioning problem*, e.g. whether the target architecture is fixed or also optimized during partitioning,

2. the supported *target architecture*, e.g. single-processor-single-ASIC or multiple-processor-multiple-ASIC architectures,

3. the *application domain*, e.g. either data flow- or control flow-dominated systems,

4. the *optimization goal* determined by the chosen *cost function*, e.g. hardware minimization under timing constraints, performance maximization under resource constraints or low-power solutions,

5. the *optimization technique*, including *heuristic* (e.g. *greedy algorithms*), *probabilistic* (e.g. *simulated annealing, genetic algorithms*) or exact methods (e.g. *integer linear programming (ILP)*), compared by *computation time* and the *quality* of the results,

6. the considered *optimization aspects*, e.g. whether communication and/or hardware sharing are taken into account or not,

7. the *granularity* of the pieces for which costs are estimated for partitioning (e.g. granules at the statement, basic block, function, process or task level),

8. the *estimation method* itself, whether the estimations are computed by special *estimation tools* or by analyzing the results of *synthesis tools* and *compilers*,

9. the considered *cost metrics* during partitioning, including cost metrics for hardware implementations (e.g. execution time, chip area, pin requirements, power consumption, testability metrics), software cost metrics (e.g. execution time, program and data memory usage) and interface cost metrics (e.g. communication time or additional resource costs),

10. the number of these considered cost metrics (e.g. whether only one hardware solution is considered for each granule or a complete AT[1] curve),

11. the degree of *automation*,

12. the degree of possible *user-interaction* to exploit the valuable experience of the designer,

13. the ability for *design space exploration* enabling the designer to compare different partitions and to find alternative solutions for different objective functions in short computation time.

The ideal hardware/software tool produces automatically a set of high-quality partitions in short, predictable computation time allowing the designer to interact.

This chapter will present the hardware/software partitioning problem as follows: First, the main aspects which have to be considered in hardware/software partitioning will be presented in section 4.1. Then, an overview of related work in the field of hardware/software partitioning is given in section 4.2. The partitioning approach in COOL is presented in section 4.3. After this, the estimation

[1] AT: Area/Time

approach integrated in COOL will be illustrated in section 4.4. In section 4.5, the generation of the partitioning graph will be described and the hardware/software partitioning problem will be formulated in section 4.6. The problem itself will be solved either using mixed integer linear programming (MILP) or genetic algorithms (GA). These optimization techniques will be introduced in section 4.7. Finally, the MILP approach will be presented in section 4.8, a combination of this MILP approach with a heuristic in section 4.9 and the GA approach in section 4.10.

4.1 The Hardware/Software Partitioning Problem

Hardware/software partitioning is a complex optimization problem which is very often defined in different ways. Before defining a precise mathematical model in section 4.6, the partitioning aspects will be described informally. The following example illustrates the hardware/software partitioning problem.

Example 3:

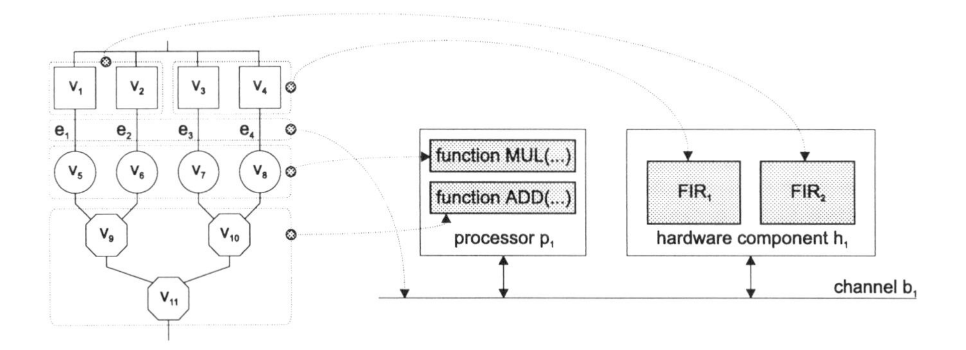

Figure 4.1. Hardware/software partitioning

In figure 4.1, a possible hardware/software partition of the 4-band equalizer is given. Two FIR-filters (v_1, v_2) are mapped to a first hardware instance ($\texttt{FIR}_1$) on ASIC h_1. Two other FIR-filters (v_3, v_4) are mapped to a second hardware instance ($\texttt{FIR}_2$). The multiplications ($v_5, \ldots, v_8$) are implemented by a function **MUL** on processor p_1. Function **ADD** implements the additions ($v_9, \ldots, v_{11}$) on p_1. The results of the **FIR** filters are calculated on h_1 and have to be transported to p_1. Therefore, the edges $e_1, \ldots, e_4$ implement interfaces using communication channel b_1. In summary, two instances of an **FIR** filter are implemented by hardware instances on h_1, and two functions (**MUL**, **ADD**) are implemented in software on p_1.

To model the hardware/software partitioning problem, the following aspects of the problem have to be considered:

- *hardware/software mapping,*
- *hardware sharing,*

- *interfacing,*
- *scheduling,* and
- *functional pipelining.*

One key problem is that most of these aspects are interacting, e.g. hardware sharing and scheduling. Two functions which share the same hardware resources have to be sequentialized. If they do not share the same hardware resources they may be executed concurrently. In the following sections, these aspects will be described in more detail.

4.1.1 Hardware/Software Mapping

The first aspect considers the decision on which processing unit an instance v_i is executed (see figure 4.2).

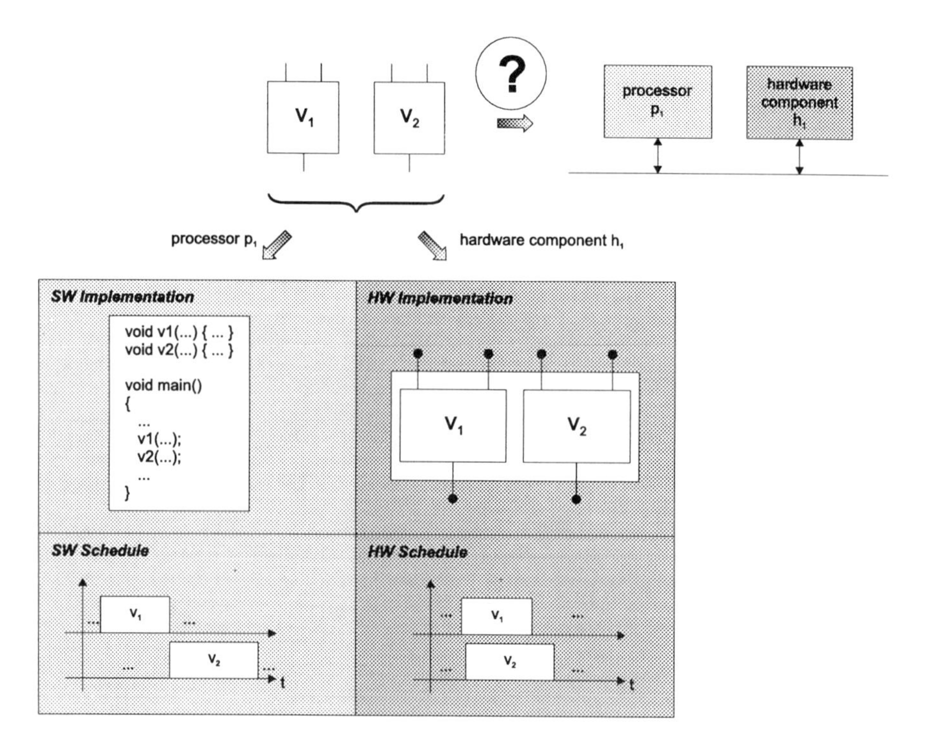

Figure 4.2. Hardware/software mapping

The key issue of this decision is that a processor can only execute a single function at a time, but hardware components will be synthesized enabling that multiple functions can be executed concurrently. As a consequence, if node v_1 is mapped to the processor, node v_2 has to be executed afterwards. In contrast, if

v_1 is mapped to a hardware component, then it might be possible to execute v_2 concurrently. Clearly, mapping and scheduling aspects depend on each other.

4.1.2 *Hardware Sharing*

One of the most important goals in hardware/software partitioning is to minimize the amount of required hardware area. Therefore, it is important to share hardware resources for different functions if possible.

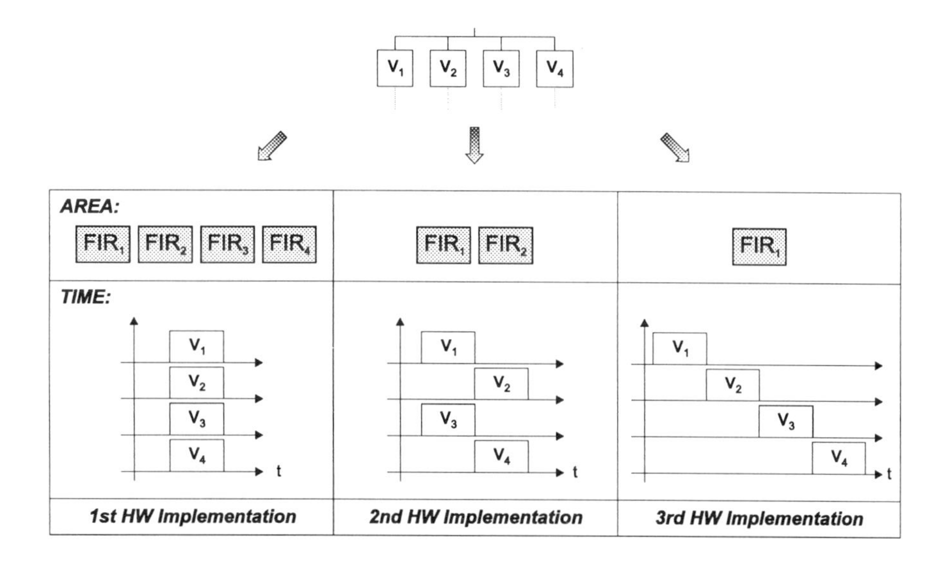

Figure 4.3. Hardware sharing

In figure 4.3, three alternative implementations of four FIR-filters $v_1, \ldots, v_4$ are depicted. The hardware costs for implementing an FIR-filter include the execution time t_{fir} and the required hardware area a_{fir}. The first implementation requires four hardware instances of an FIR-filter, but all of them can be executed concurrently. Hence, the total hardware area is $A \approx 4 * a_{fir}$ and the total execution time is $T \approx t_{fir}$. The third implementation requires only one hardware instance, but in such a case all functions have to be sequentialized ($A \approx a_{fir}, T \approx 4 * t_{fir}$). The second implementation represents a compromise between performance and chip area. It requires two hardware instances of an FIR-filter, but instances v_1, v_2 and v_3, v_4 have to be sequentialized ($A \approx 2 * a_{fir}, T \approx 2 * t_{fir}$). Summarizing, hardware sharing and scheduling are also interacting aspects of the partitioning problem.

4.1.3 Interfacing

As described before, the mapping determines on which processing unit an instance of a system component should be executed. If two instances, which exchange data, are mapped to different processing units, then additional *communication* is required to implement these data transfers.

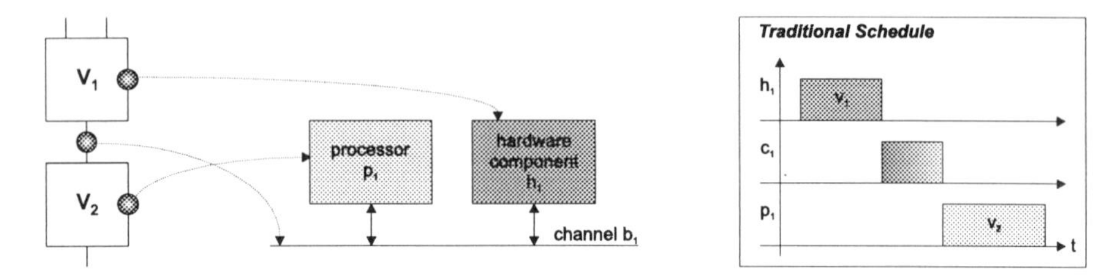

Figure 4.4. Interfacing

In figure 4.4, instance v_1 is mapped to hardware component h_1 and instance v_2 to processor p_1. The data computed by v_1 has to be transported to v_2. For this reason, a communication channel has to be allocated to implement the data transfer (in figure 4.4 this is very simple, because only channel b_1 is available). Furthermore, the communication event has to be scheduled on the allocated communication channel. This means that an additional communication time has to be considered which is required to transport the data using the selected channel.

4.1.4 Scheduling

The scheduling aspects have been described indirectly in the paragraphs before. The task of scheduling is always to prevent *resource conflicts*. The model describing the scheduling constraints has to guarantee that a processor is only occupied by one function at a time. The execution of functions sharing the same hardware resources also have to be scheduled. Finally, all transfers on each communication channel have to be sequentialized as described in section 4.1.3.

4.1.5 Functional Pipelining

The overall system execution time of the schedule can be optimized by using *functional pipelining*. The idea is, to restart the system for new incoming data, although it has not finished execution for the old inputs. An example of functional pipelining is given in figure 4.5.

Functions v_1 and v_2 are executed on the same processor p_1 whereas functions v_3 and v_4 are executed on the same hardware component h_1. The system

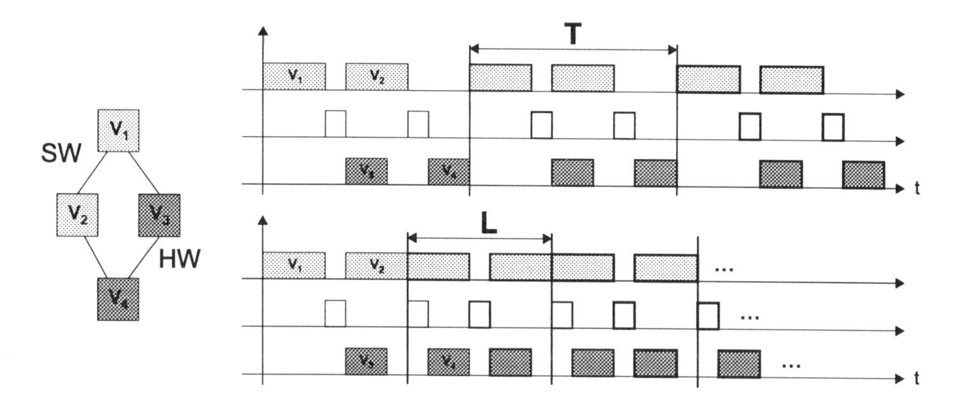

Figure 4.5. Functional pipelining

computes the result for the given inputs and then it is restarted. The *total execution time* of the system is T. Using functional pipelining, the system can be restarted earlier. In this case, the system execution time is reduced to L which represents the *latency* of the pipeline.

4.2 Related Work

In the literature, a variety of hardware/software partitioning approaches has been published. Therefore, only the most important ones will be presented.

One of the first partitioning approaches has been developed for the COSYMA system. Ernst has presented an approach [EHB93], where hardware/software partitioning is based on *simulated annealing* using estimated costs. The partitioning algorithm is *software-oriented*, because it starts with a first non-feasible solution consisting only of software components. In an *inner loop partitioning* software parts of the system are iteratively realized in hardware until all timing constraints are fulfilled. To handle discrepancies between estimated and real execution time, an *outer loop partitioning* restarts the inner loop partitioning with adapted costs [HHE94]. The outer loop partitioning is repeated until all performance constraints are fulfilled.

Another pioneering hardware/software partitioning approach has been presented by Gupta and De Micheli [GuMi92, GuMi93] realized in the VULCAN system [GCM92]. Their approach is *hardware-oriented.* It starts with a complete hardware solution and iteratively moves parts of the system to the software as long as the performance constraints are fulfilled. In this approach performance satisfiability is not part of the cost function. For this reason, the algorithm can easily be trapped in a local minimum. After computing a hardware/software partition, the hardware component can be partitioned onto a set of ASICs as described in [GuMi90].

Vahid has developed several hardware/software partitioning algorithms. In [VGG94] he uses a relaxed cost function to satisfy performance in an inner partitioning loop and to handle hardware minimization in an outer loop. The cost function consists of a very heavily weighted term for performance and a second term for minimizing hardware. The authors present a *binary-constraint search algorithm* which determines the smallest size constraint (by binary search) for which a performance satisfactory solution can be found. The partitioning algorithm minimizes hardware, but not execution time. Another approach published by Vahid has extended the well-known *Kernighan/Lin heuristic* for hardware/software partitioning [Vahi97, VaLe97]. The key extension customizes the heuristic and data structure to rapidly compute execution-time and communication metrics. The goal is to minimize the execution time while satisfying resource constraints for a single-processor-single-ASIC target architecture.

Kalavade and Lee [KaLe94] present an algorithm called GCLP[2] that determines for each node iteratively the mapping to hardware or software. The GCLP algorithm does not use a hardwired objective function, but it selects an appropriate objective according a global time-criticality measure and another measure for local optimum. The results are close to optimum and the computation time quadratically grows to the number of nodes. This approach has been extended to solve the *extended partitioning problem* [KaLe95, KaLe97] including the *implementation selection problem*, where not only one hardware solution for each node is considered by the algorithm, but a complete AT-curve. The main disadvantage of this approach is that additional interface costs are not estimated but considered as being constant. Thus, communication cost may be over- or under-estimated.

Knudsen [Knud95, KnMa96] has developed an algorithm called PACE. PACE is integrated in the LYCOS [MGK+97] environment. It solves both the problem of hardware minimization under timing constraints and the problem of performance maximization for a restricted amount of hardware area. The algorithm is based on *dynamic programming* and supports a single-processor-single-ASIC target architecture. It has been used in particular for design space exploration.

Teich and Blickle [TBT97, Blic97] present an *evolutionary approach* to system-level synthesis optimizing the mapping of a coarse-grain data flow graph-based specification onto a heterogeneous target architecture consisting of multiple processors and ASICs. The advantage of this approach is that the target architecture is not fixed. The evolutionary approach performs the selection of a target architecture (among a specified set of possible architectures) and the mapping of the specification onto the target architecture in a single optimization run.

[2]GCLP: Global Criticality / Local Phase

Author	Domain	Arch.	Algorithm	Objective	Costs	Est. Tool
Ernst	control	1,1,1	simulated ann.	SW acc.	1,1,1	yes
Gupta	both	1,n[3],1	greedy	HW min.	1,1,1	yes
Vahid	both	1,1,1	binary search	HW min.	1,1,-	yes
Vahid	both	1,1,1	Kernighan/Lin	Time min.	1,1,1	yes
Kalavade	data	1,1,1	constructive	HW min.	1,1,c[4]	yes
Kalavade	data	1,1,1	constructive	HW min.	1,n,c	yes
Knudsen	control	1,1,1	dynamic prog.	HW/Time min.	1,1,1	yes
Teich	data	n,n,n	evolutionary	user-defined	1,1,1	yes

Table 4.1: Comparison of different partitioning approaches

In table 4.1, some characteristics of the presented hardware/software partitioning approaches are summarized comparing

- the application domain,
- the supported target architecture n_p, n_a, n_c consisting of n_p processors, n_a ASICs, and n_c communication channels,
- the partitioning algorithm used to solve the problem,
- the optimization goal,
- the numbers n'_p, n'_a, n'_c of considered cost estimations for each partitioning granule for software (n'_p), hardware (n'_a) and communication (n'_c) implementations,
- the estimation method.

This overview is not complete, because there is a variety of other automatic partitioning approaches based either on *clustering algorithms* [BRX94], [BFS93], on *MILP* [Bend96], [NiMa97], on *simulated annealing* [PeKu93, EPKD97], [Axel97] on *tabu search* [EPKD97], [Axel97], on *dynamic programming* [JEO+94] or on *genetic algorithms* [Axel97], [SMB97].

4.3 Hardware/Software Partitioning in COOL

The hardware/software partitioning approach implemented in COOL works as illustrated in figure 4.6. First, the designer selects the structural system to be partitioned from the *system library* while determining a *target architecture* onto which the system should be partitioned. After the *design constraints* have been specified, the partitioning phase starts with computing a *partitioning graph* for the system specification. During this step, the nodes are weighted with costs for implementing them using components of the chosen target architecture. These costs have been computed in a preprocessing step and stored in the *cost library*.

[3]Each hardware part is partitioned onto n ASICs after performing hardware/software partitioning.

[4]Communication costs are estimated by a constant value.

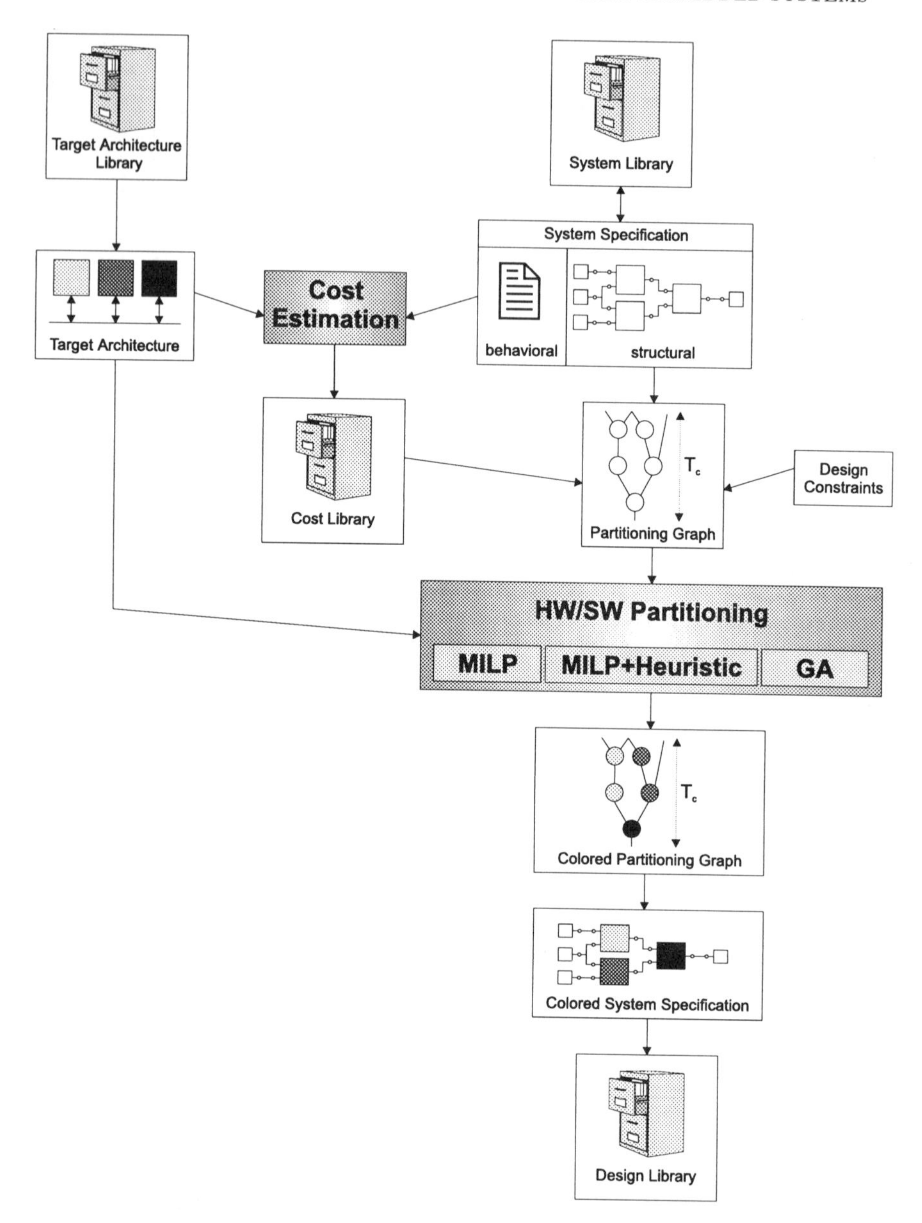

Figure 4.6. Hardware/software partitioning in COOL

In addition, the design constraints are annotated to the partitioning graph. Then, the designer has to choose one of three available algorithms for solving the hardware/software partitioning problem. The result of all algorithms is a *colored partitioning graph* annotated with *schedule* times for the nodes. Finally, the results are copied back to the system specification and stored in a *design library*.

Compared to related hardware/software partitioning approaches, COOL differs mainly in two aspects:

1. The cost estimation technique uses no special estimation tools. In COOL *static analysis* methods are applied to the results of high-level synthesis and software compilation to obtain very accurate results. The reason for this is that COOL tries to support data flow dominated systems having hard timing constraints. Therefore, the worst-case execution times of hardware and software implementations have to be determined very precisely. The disadvantage of an increased computation time to calculate these values for the cost metrics is compensated by a higher precision. Naturally, this may lead to fewer design iteration steps.

2. COOL uses different optimization approaches to solve the hardware/software partitioning problem. Three new fully-automatic algorithms to solve the partitioning problem will be presented, supporting multi-processor-multi-ASIC target architectures. All of them have been integrated in the COOL framework and allow intensive user-interaction.

   (a) The first approach is based on *mixed integer linear programming (MILP)* to solve the partitioning problem. A formulation of the MILP model will be introduced in detail in section 4.8. This approach produces optimal results, but the drawback of solving MILP models is often a high computation time.

   (b) To reduce the computation time, a second algorithm using MILP has been developed which splits the partitioning problem in two subproblems. In a first phase, a mapping of nodes to hardware or software is calculated by estimating the schedule times for each node with heuristics. During the second phase a correct schedule is calculated for the resulting HW/SW mapping of the first phase. It will be shown that this *heuristic scheduling* approach strongly reduces the computation time while the results are nearly optimal for the chosen objective function. Nevertheless, the computation time will explode for complex systems in some cases.

   (c) Both approaches described before (and the first one in particular) suffer from the fact that the computation time to produce a solution is not predictable. Therefore, their application for design space exploration is of limited benefit. The third approach is based on *genetic algorithms* producing good results in predictable time. This approach is very well suited to be utilized for design space exploration.

Compared to the partitioning approaches summarized in table 4.1, COOL can be classified as depicted in table 4.2.

Domain	Arch.	Algorithm	Objective	Costs	Estimation
...	...	...	...	...	...
data flow	n,n,n	MILP	user-defined	1,1,1	Synthesis/
data flow	n,n,n	MILP + heuristic	user-defined	1,1,1	Compilation
data flow	n,n,n	GA	user-defined	n,n,1	Tools

Table 4.2. Classification of hw/sw partitioning approaches in COOL

4.3.1 Special Aspects of HW/SW Partitioning in COOL

The model for hardware/software partitioning in COOL considers the aspects presented in section 4.1: *hardware/software mapping*, *hardware sharing*, *interfacing*, *scheduling*, and *functional pipelining.* These aspects will be modelled with the help of

- variables and constraints in the MILP approach and
- encoding restrictions and the fitness function in the GA approach.

Some general comments to special issues of the hardware/software partitioning model in COOL will be described in the following. These comments are concerned with a precise interfacing and scheduling model.

4.3.1.1 Precise Interfacing Model. In most approaches found in the literature, a data transfer is modelled by one additional communication block (see section 4.1.3) consuming some communication time as depicted in figure 4.7.

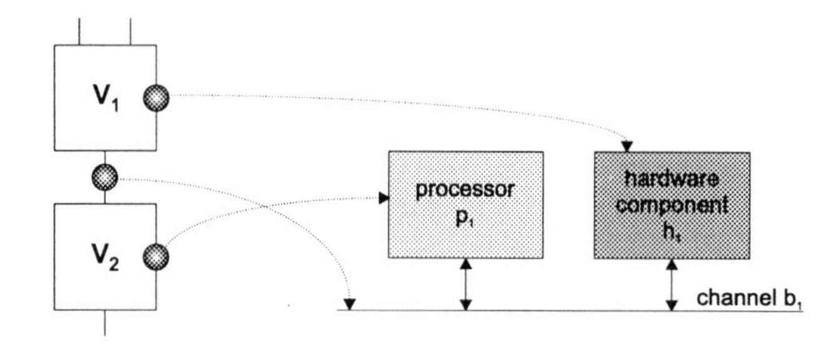

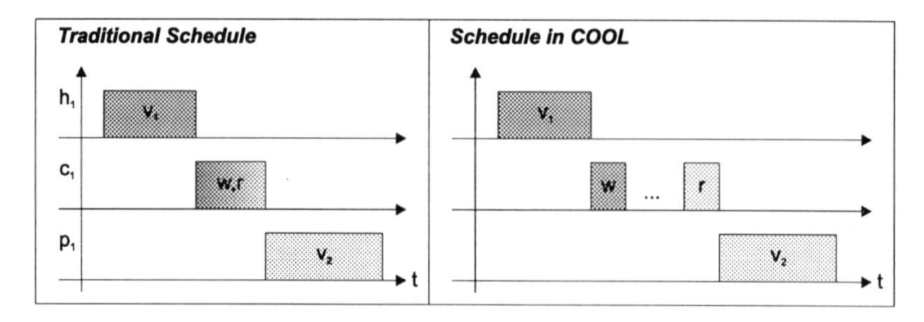

Figure 4.7. Communication modelling in COOL

In COOL the interfacing aspect is modelled more precisely, because a data transfer is divided in a WRITE- and READ-phase which may be decoupled. This allows a very precise modelling of different communication alternatives implementing these interfaces between processing units (see figure 4.8).

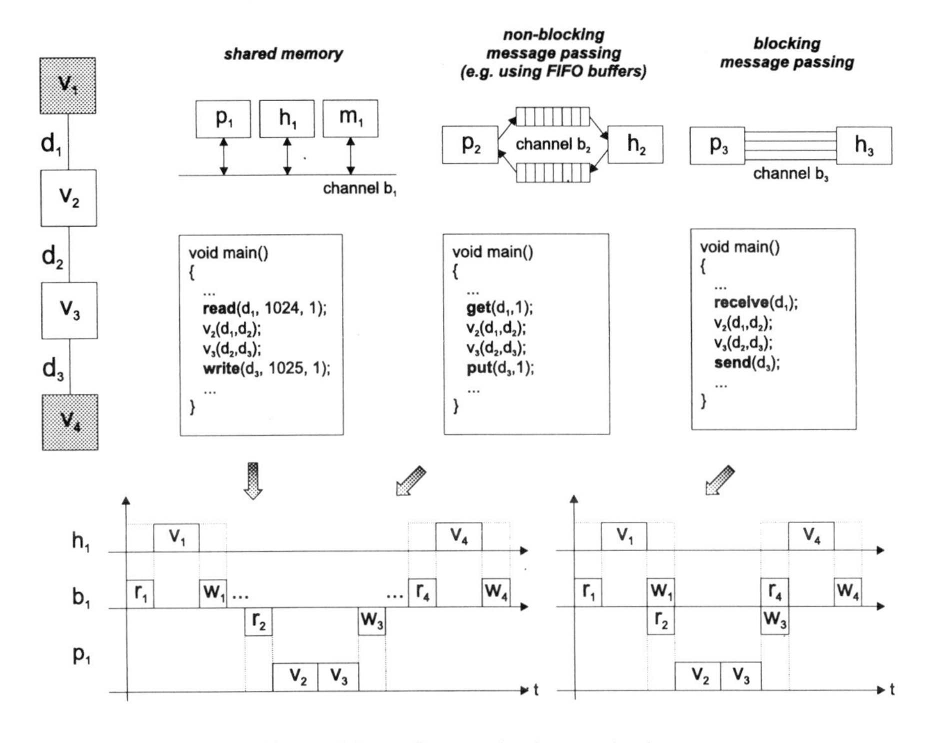

Figure 4.8. Communication mechanisms

The interface can be implemented, for example, by using a *shared memory* or *message passing communication* mechanisms which will be described in more detail in chapter 5. They have in common that data is read or written using a *communication channel* with thc hclp of abstract communication routines. These abstract communication routines are called READ (read, receive, or get in figure 4.8) and WRITE (write, send, or put) in the following. The difference in modelling the communication schemes is that for *blocking message passing* the WRITE- and READ-operations occur simultaneously. In this case, the writing unit (e.g. p_3) blocks until the other processing unit (e.g. h_3) is ready for receiving the data. In contrast, when using shared memory or *non-blocking message passing* communication, the processing units do not have to be synchronized. The writing unit writes the data into external memory (or the *FIFO*[5] buffer) and the receiving unit reads the data from this storage whenever it is ready.

[5]FIFO: First In First Out

Clearly, for all communication schemes the `READ`- and `WRITE`-operations have to be scheduled on the communication channel to prevent conflicts. These additional interface constraints increase the complexity of the scheduling problem drastically.

4.3.1.2 Scheduling Restrictions.

The scheduling aspects have been described in section 4.1.4. In the following, two additional scheduling restrictions in COOL concerning

1. a correct model of channel accesses, and
2. the execution order of instances mapped to processors or ASICs

will be presented. These restrictions support a correct partitioning model considering aspects of the final implementation during co-synthesis.

Channel Accesses. It is important to mention that if a processor reads input data or writes output data using the communication channel, it is not able to compute another function simultaneously. This overlap has to be prevented by additional scheduling constraints. In figure 4.9, an example of an invalid schedule is given.

Example 4:

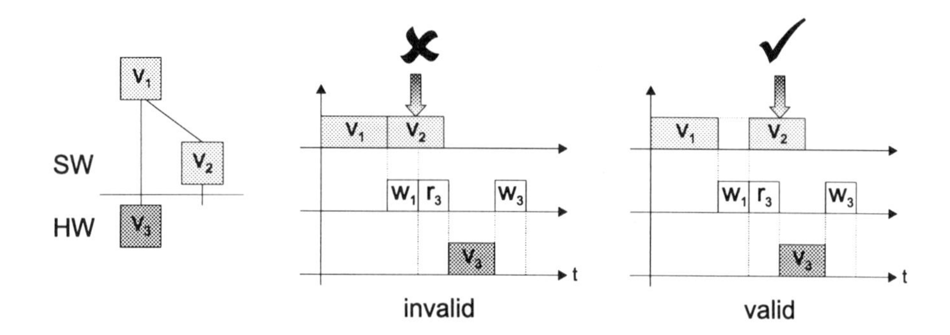

Figure 4.9. No computation during channel access

Functions v_1 and v_2 have been mapped to the same processor. Function v_2 starts executing directly after v_1 has finished. In parallel the `WRITE`-operation of v_1 is scheduled on the communication channel to transport the computed data to v_3. This schedule is invalid, because the processor is also occupied when function v_1 writes its results. In the valid schedule, v_2 is started after v_1 has written its results.

Execution Order. After partitioning, the system will be implemented by a hardware/software system controlled by a system controller realizing a *run-time scheduler.* This run-time scheduler activates a processing unit (processor or hardware component) to execute a special function v_i if

1. all data for v_i is available and
2. the processing unit is not occupied.

If the processing unit has been activated by the system controller, it works corresponding to the following *execution order*:

1. First, the necessary input data for function v_i computed on other processing units is read using the communication channel.
2. After all input data has been read, the function v_i is executed.
3. Finally, the computed results of v_i which are required by functions implemented on other processing units are written using the communication channel.

After all results have been written, the processing unit sends a message to the system controller to signify the end of computation for v_i. For this reason, the processing units implementing systems designed with COOL work in a *non-preemptive* mode:

> The execution order (READ→EXECUTE→WRITE) for executing a function f_1 on a processing unit P will not be interrupted by the execution of another function f_2 (or their R/W accesses) on P.

In figure 4.10, an example of a violation of this execution order is depicted.

Example 5:

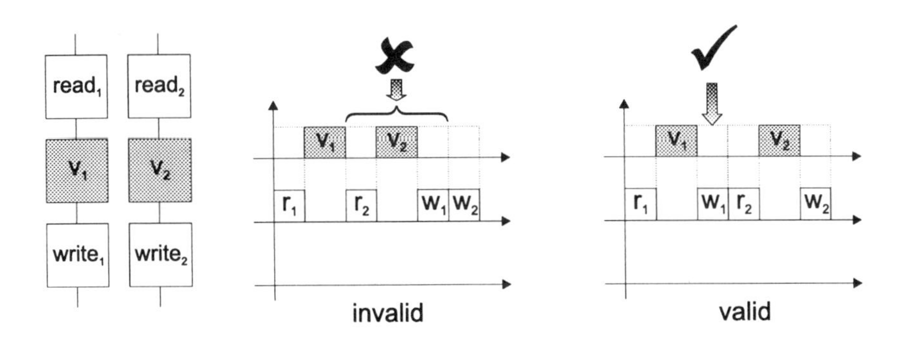

Figure 4.10. Violation of execution order

Functions v_1 and v_2 have been mapped to the same processing unit. The first schedule is invalid, because v_2 starts reading input data although v_1 has not written its result. Therefore, the execution order READ→EXECUTE→WRITE for v_1 is interrupted by v_2. The correct schedule is that v_2 starts reading when v_1 has written its result.

4.4 Hardware/Software Cost Estimation

Many co-design approaches, e.g. [BFS95, CaWi96, CEG+96, EHB93, GuMi96, GVN94, MGK+97], use special estimation tools to estimate hardware and software costs. Most of these approaches use *dynamic analysis* in the form of *profiling* where the specification is simulated for a set of stimuli. Profiling allows the computation of some *average execution time* for the simulated system specification. However, hard timing constraints cannot be guaranteed, because the computation of the *worst-case execution time* cannot be determined by profiling. To support hard timing constraints, the quality of the estimations is very important, because inaccurate estimations very often lead to designs violating some performance constraints.

The main objective of the hardware/software partitioning approach realized in COOL is minimizing hardware area under timing constraints. Therefore, in contrast to most approaches, COOL uses no special estimation tools for cost estimation. Instead, the results of hardware high-level synthesis and software compilation are analyzed by *static analysis*. Figure 4.11 gives an overview of the cost estimation approach integrated in COOL.

Each behavioral system specification stored in the system library is compiled into an internal syntax graph model using the commercial `VHDL` frontend VTIP[6] [Clsi90]. Then, software source code (`C` or `DFL`) and synthesizable[7] hardware source code (`VHDL`) is generated. For each processor of the *target architecture library* software parts are compiled using a standard `ANSI-C` compiler. In addition, work has been started to integrate the *retargetable code generator* RECORD [Leup97] in COOL. Using RECORD, COOL will also able to support target architectures containing ASIPs or other processors for which no other compiler exist. In this case, a processor model has to be specified with which RECORD is able to generate code for a `DFL` specification. On the hardware side, for each ASIC technology stored in the target architecture library the hardware parts are synthesized by the *high-level synthesis* tool OSCAR [LMD94, Land98]. The technology knowledge is defined in the *component libraries* of OSCAR.

The structural systems consist of behavioral systems, for which the costs have already been estimated. Therefore, only the costs for additional communication, which might occur on wires, have to be estimated. For this reason, for each data type transported on a wire of the structural system the communication time is estimated for each communication channel.

The disadvantage of this estimation approach is an increased computation time caused in particular by using high-level synthesis for hardware cost estimation. To reduce this disadvantage, a cost library is integrated in COOL allowing to store the results of compilation and synthesis. Therefore, for each behavioral

[6] VTIP: `VHDL` Tool Integration Platform

[7] The `VHDL` system specification is simulatable but not synthesizable.

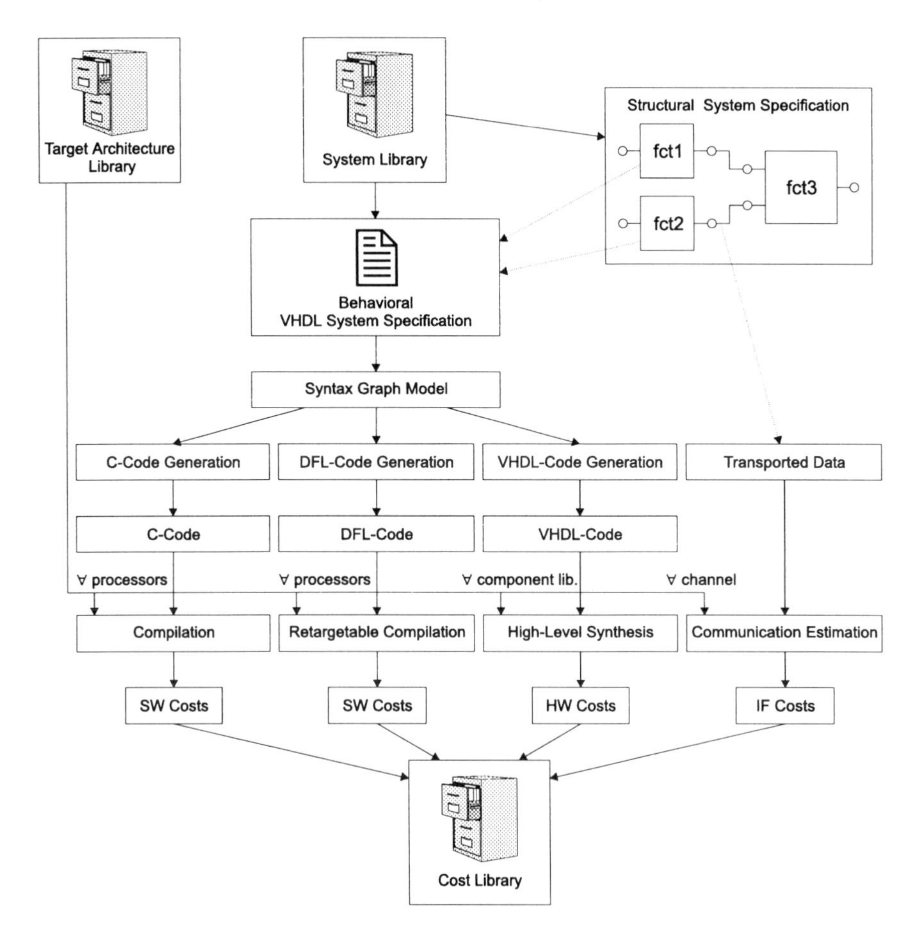

Figure 4.11. Cost estimation flow in COOL

system the values for the cost metrics are only estimated once and stored in the *cost library*. Afterwards, the costs can be read from this library. The advantage of this approach is an increased accuracy of the estimations probably leading to fewer design iterations.

As mentioned before, to support hard timing constraints, the worst-case execution time has to be estimated for hardware and software implementations by static analysis. As a consequence of performing static analysis, *non-deterministic* constructs in the specification, like *unbounded loops*, have to be transformed into constructs with well-defined bounds (see figure 4.12).

Otherwise, the worst case timing behavior of a certain function could not be estimated precisely. Hence, hard timing constraints could only be guaranteed with a certain probability.

```
...
while i < max loop                  for i=1 to 10 loop
                                      if i < max then
  ...                                   ...
                                      end if;
end loop;                           end loop;
...
```

Figure 4.12. Transformation of an unbounded loop

The following cost metrics are estimated and considered during hardware/software partitioning in COOL:

1. *software cost metrics*:
   - software program memory usage,
   - software data memory usage,
   - software execution time,
2. *hardware cost metrics*:
   - hardware area,
   - hardware execution time,
3. *interface cost metrics*:
   - communication time.

It should be mentioned that other cost metrics, e.g. power consumption, pin requirements or testability metrics could easily be considered additionally when special estimation tools would be integrated. But in this book only the cost metrics presented above will be considered. In the following, the methods to estimate these software, hardware and interface cost metrics are described in sections 4.4.1-4.4.3.

4.4.1 Software Cost Estimation

Software costs are computed as illustrated in figure 4.13. First, the generated software specification is compiled using an ANSI-C compiler for the chosen processor. The result of this compilation step is an *assembler program* which is analyzed to compute the values for the cost metrics.

The required *program memory* is estimated by the number of instructions of the generated assembler code. The amount of *data memory* is estimated by accumulating the memory requirements of all variables of the C code. The *software execution time* cost metric represents the *worst-case execution time* of the generated assembler code. Static analysis is applied to identify the *critical path* in the code using an *instruction set description* of the processor. This

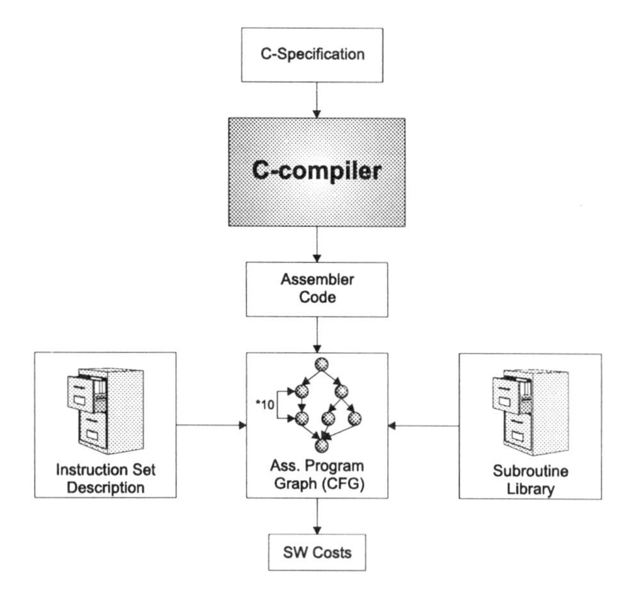

Figure 4.13. Computation flow for software cost estimation

description contains the number of cycles required to execute each instruction of the instruction set. In addition, a *subroutine library* is necessary for computing the execution times of subroutines (such as arithmetic functions). The following example illustrates the algorithm executing the static analysis phase.

Example 6:

In figure 4.14, assembler code generated for a MOTOROLA DSP56001 is depicted. The code contains three different routines. The `IRQ_Service_Routine` represents the top-level routine, calling other functions, including `read_signed_int_24` (reading a 24 bit signed integer number from external memory) or `fct` (computing a certain function). The worst-case execution time is estimated in five steps.

1. First, a control flow graph is constructed, where each node represents an assembler instruction (including arithmetic operations, branches, loops or subroutine calls) or a special node for storing information (Function Begin, Function End). The edges represent the control flow of the assembler code, including edges for branches and loops.
2. In a second step, all nodes representing instructions are annotated with the number of required *instruction cycles* stored in the instruction set description. In addition, the nodes representing subroutine calls are annotated with the number of required instruction cycles, to execute the subroutine. These numbers are defined in the subroutine library. Edges, representing loops, are weighted with the number of loop iterations scanned from the corresponding loop-instruction (remember: only bounded loops are allowed!).
3. Then, for all backward-edges (v_i, v_j) representing loops: for all nodes, being element of path (v_i, v_j), the number of instruction cycles is multiplied by the number of loop iterations.
4. Finally, the worst case execution time is computed by accumulating the instruction cycles of the nodes. This is done using a graph traversal, starting from the

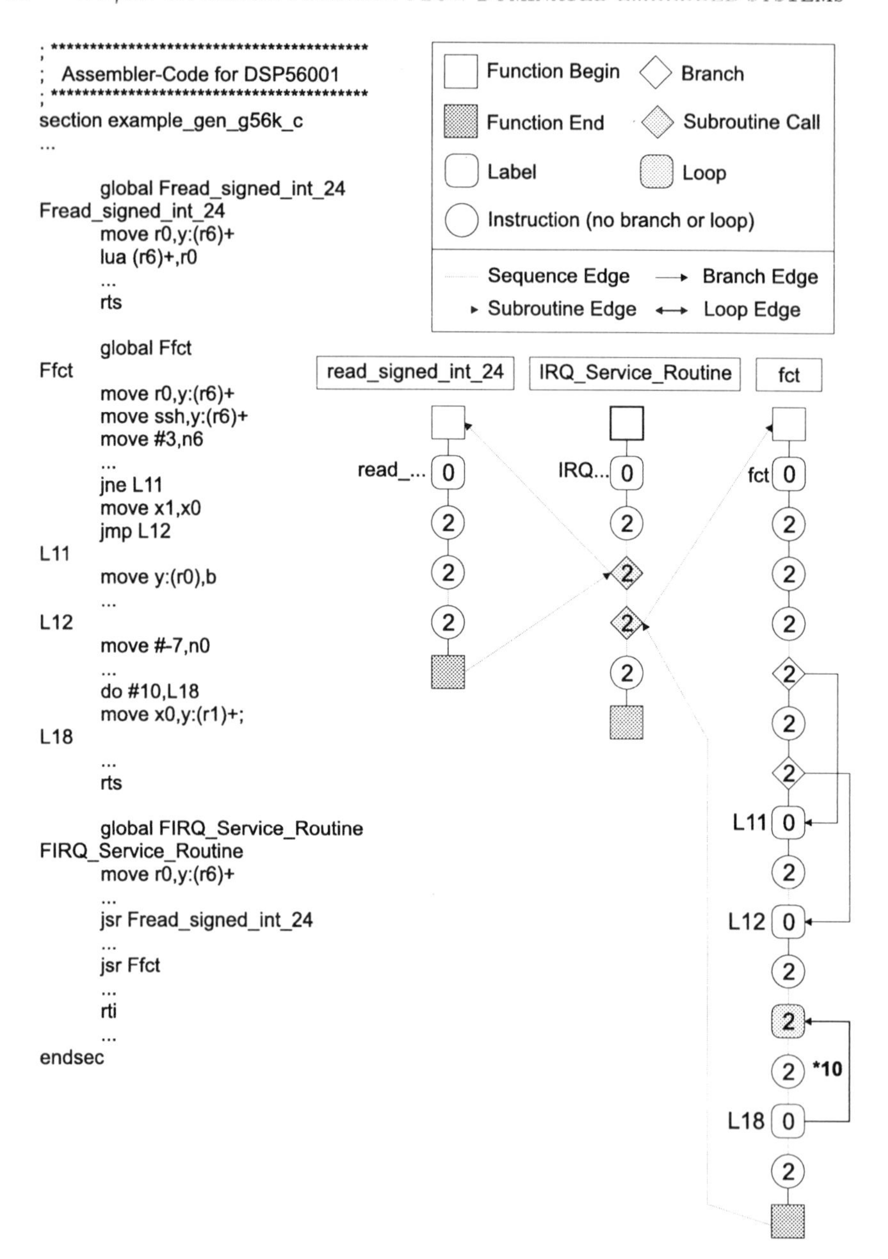

Figure 4.14. Software execution time estimation

'Function Begin'-node of the `IRQ_Service_Routine`. If a branch node is reached, then the path with the maximum number of instruction cycles is considered.

5. After the number N of *instruction cycles* for the critical path has been determined, the worst-case execution time is computed by multiplying N by the *instruction cycle time* of the target processor.

4.4.2 Hardware Cost Estimation

The estimation phase determining hardware costs uses the *high-level synthesis* tool OSCAR[8] and its backend tool, called OSBACK[9]. The purpose of high-level synthesis is to map operations of the behavioral specification to certain control steps and hardware resources. Therefore, the high-level synthesis problem can be divided into three different tasks:

1. *Scheduling*: determines at which control step an operation is executed.
2. *Allocation*: determines which hardware components are used to execute operations.
3. *Binding*: denotes the assignment of an operation to an instance of a hardware component.

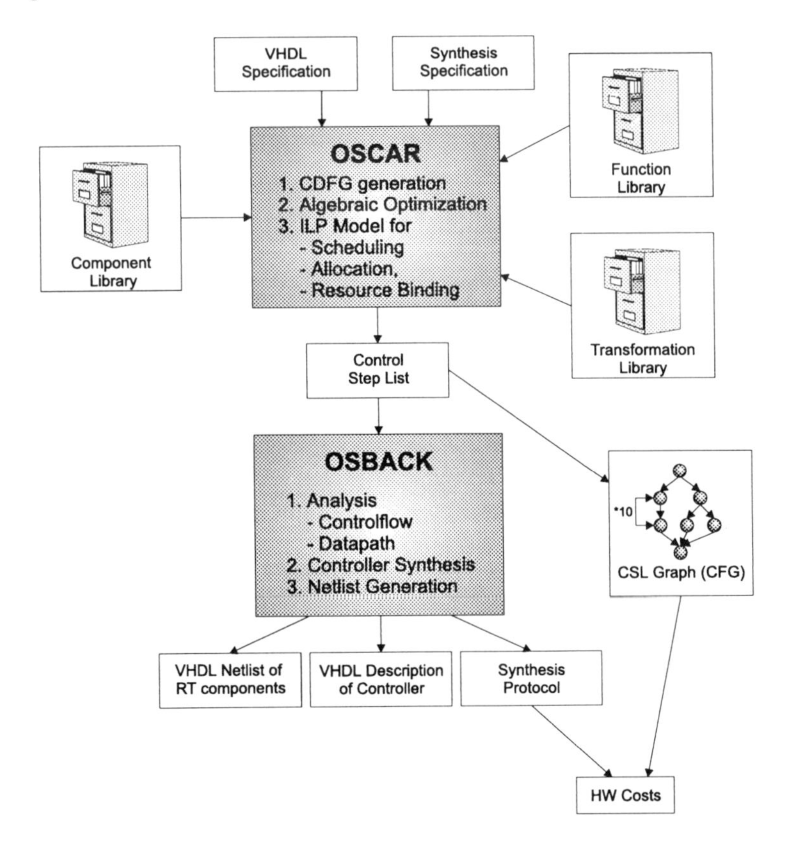

Figure 4.15. Design flow for hardware cost estimation

[8]Oscar: Optimum Simultaneous Scheduling, Allocation and Resource Binding
[9]OsBack: Oscar Backend

The OSCAR synthesis system [LaMa93, LMD94, Döme94, LaMa97, Land98] is a high-level synthesis tool performing the scheduling, allocation and binding tasks simultaneously using *integer linear programming (ILP)*. The basic variables of the generated ILP model are triple-indexed variables $x_{i,j,k}$, modelling whether an *operation* j will be started on *hardware resource instance* k at *control step* i. One main advantage of OSCAR is the support of commercial component libraries, e.g. from COMPASS [Comp91] or SYNOPSYS [Syno92]. In figure 4.15, an overview of the design flow is given to compute hardware cost estimations using OSCAR and OSBACK.

The inputs to OSCAR include a behavioral VHDL specification and a *synthesis specification file*. This file contains important information required during synthesis, e.g. the name of the CLOCK- and the RESET-signal, the *cycle time*, user-defined timings and manual bindings of operations to control steps and/or component instances, etc. In the first step, the behavioral VHDL specification[10] is read and a frontend transforms the specification into an internal representation, a combined *control/data flow graph*. This control/data flow graph represents a control flow graph where the nodes are *basic blocks*. Each basic block is a data flow graph including all operations of a sequence of assignments. These operations (functions) and their attributes (number of arguments, commutativity, associativity, the availability of neutral elements, etc.) are defined in the *function library*. In figure 4.16, the specification of the multiplication function in the function library is depicted.

```
FUNCTION          *
ARGUMENTS         2
COMMUTATIVITY     TRUE
ASSOCIATIVITY     TRUE
NEUTRALS          (IDENTITY, 1)
EXTENSIONS        SIGN, SIGN
RESULT_WIDTH      MAX(#1, #2)
```

Figure 4.16. Function description

```
CLASS AssociativityRules1 IS
  VARIABLE (a, b, c : bit_vector);
BEGIN
  (a + b) + c;
  a + (b + c);
END AssociativityRules1;
```

Figure 4.17. Transformation rule

After reading the function library, the *transformation library* is read. This library contains a set of *transformation rules*, e.g. the associativity rule as depicted in figure 4.17, defining equivalent alternatives of expressions. Using this transformation library, *algebraic transformations* are applied to compute a set of equivalent alternatives for each basic block. During the synthesis step, these algebraic transformations lead to a significant cost reduction.

The *component library* defines the technical data of the components stored in the commercial component library as depicted in figure 4.18.

The required chip area is defined by attribute `total_area`, the execution delay is defined by the **`AFTER statement`**. The link to the commercial COMPASS

[10] OSCAR supports a synthesizable subset of VHDL described in [Marm93].

```
-- Signed Multiplier, 16 Bit --------------------------------------------
COMPONENT o_smul16x16_1 OF CLASS FUNCTION_UNIT IS
 PORT(  in1      :  IN BIT_VECTOR(15 DOWNTO 0);
        in2      :  IN BIT_VECTOR(15 DOWNTO 0);
        outlo    :  OUT BIT_VECTOR(15 DOWNTO 0);
        outhi    :  OUT BIT_VECTOR(15 DOWNTO 0));
 ATTRIBUTE
        total_area       :  2298.00 * 1669.00,    -- square lambda
        costs            :  15000;                -- virtual benchmark costs
 ATTRIBUTE
        generator        :  "COMPASS Data Path Compiler",
        cell_library     :  "vdp370d",
        cell             :  "vdp3mlt004";
BEGIN
        outlo <= in1 * in2 AFTER 60.3 NS;
END o_smul16x16_1;
```

Figure 4.18. Component specification

library is specified by attributes **`cell_library`** and **`cell`**. In addition, the component specification defines which function of the function library (in figure 4.18 it is the multiplication defined by **`'in1 * in2'`**) can be implemented by which component of the component library.

```
SCHEDULE EXAMPLE IS
...
BEGIN SCHEDULE
  WAIT ON (CLK[0:0]);
  BEGIN BLOCK BLOCK_1
    1 REG_1 ...                    -- control step 1
    2 O_CMP32X32_1_1 ...           -- control step 2
    2 COND_4 ...                   -- control step 2
  END BLOCK;
  IF (COND_4[0:0]) THEN
     BEGIN BLOCK BLOCK_2          ...  END BLOCK;
     IF (COND_3[0:0]) THEN
        BEGIN BLOCK BLOCK_3       ...  END BLOCK;
        IF (COND_1[0:0]) THEN
           BEGIN BLOCK BLOCK_4    ...  END BLOCK;
        ELSE
           BEGIN BLOCK BLOCK_5    ...  END BLOCK;
        END IF;
        BEGIN BLOCK BLOCK_6       ...  END BLOCK;
        WHILE (COND_2[0:0]) LOOP
           BEGIN BLOCK BLOCK_7    ...  END BLOCK;
        END LOOP;
     END IF;
  ELSE
     BEGIN BLOCK BLOCK_9          ...  END BLOCK;
  END IF;
END SCHEDULE;
```

Figure 4.19. Control step list

Using all information stored in the CDFG, the synthesis specification file and the libraries, an ILP model is generated including scheduling, allocation and binding constraints. This ILP problem is solved using an MILP solver either LP_SOLVE [Berk93] or OSL [OSL92]. The solution (the complete set of bounded $x_{i,j,k}$ variables) is transformed into a *control step list* (see [Döme94]) which is a textual control flow description containing basic blocks, as depicted in figure 4.19.

The backend OSBACK [Kone95] represents the interface between OSCAR and commercial synthesis tools like SYNOPSYS or COMPASS. OSBACK reads this control step list and produces a controller and a netlist of instantiated RT-components, both specified in VHDL.

In COOL, the results of OSBACK are used for estimating the hardware resource costs. OSBACK produces a detailed synthesis protocol including the number and area of

- ports (in, out, inout, buffer),
- registers,
- functional units, and
- multiplexers.

The total *hardware area* is estimated by accumulating the area of the parts described before. However, only the costs for the data path are considered in this step. The costs for the controller generated by OSBACK for this data path are not considered. To be more precise, the controller costs have to be estimated additionally, but for data flow dominated systems the required chip area for the controller is much smaller compared to the chip area used for the data path. Therefore, controller costs are not estimated during chip area analysis in COOL.

The *hardware execution time* is estimated by static analysis of the control step list. The control step list is transformed into a control flow graph, where each node represents a basic block. A node is weighted with the number of control steps required to execute its basic block. This information can be read from the control step list. For example, `block_1` in figure 4.19 requires two control steps (see comment statements in figure 4.19). The edges of the generated graph represent control flow edges including branches and loops. The structure of this graph and the method for computing the worst-case execution time is very similar to the approach for estimating the software execution time described in section 4.4.1. Therefore, it will not be described in detail.

4.4.3 Interface Cost Estimation

Additional interface costs have to be taken into account if data has to be transported from one processing unit to another. Using *shared memory communication*, data D consisting of N_D bits, has to be written into memory using a data bus db with a bit width of N_{db} bits. The number of clock cycles for reading (writing) a word is defined as $N_{read}(N_{write})$ in the *target architecture library*. For a cycle time t_{cycle}, the communication time $c^{tr}(D, db)$ for reading and the time $c^{tw}(D, db)$ for writing data D using data bus db can be estimated by equations 4.1 and 4.2.

$$c^{tr}(D, db) = \lceil N_D/N_{db} \rceil * N_{read} * t_{cycle} \quad (4.1)$$

$$c^{tw}(D, db) = \lceil N_D/N_{db} \rceil * N_{write} * t_{cycle} \quad (4.2)$$

4.5 Generation of the Partitioning Graph

The *partitioning graph* is the basic data structure for hardware/software partitioning. It has to represent the flattened system specification which should be mapped to a certain target technology. Therefore, it consists of

- a set of *computation nodes* representing instances of different system components and
- a set of *communication nodes* representing `READ-/WRITE`-accesses to model additional communication. This communication might be necessary for implementing a hardware/software system.

The formal definition of the partitioning graph is as follows:

Definition 1: *Partitioning graph*

A **partitioning graph** G^P *is defined as a 4-tuple*

$$G^P = (V, E, C, I).$$

G^P *is a directed graph* (V, E). *The set of nodes* $V = \{v_1, \ldots, v_{n_V}\}$ *consists of two disjoint sets* $V = V^I \cup V^{RW}$:

- *a set* $V^I = \{v_{i_1}, \ldots, v_{n_{V^I}}\}$ *of computation nodes (representing instances of system components* $c \in C$*) and*
- *a set* $V^{RW} = \{v_{i_1}, \ldots, v_{n_{V^{RW}}}\}$ *of communication nodes representing R/W-accesses to a communication channel.*

The set of communication nodes V^{RW} *itself is divided into four disjoint sets*

$$V^{RW} = V^{RI} \cup V^{WO} \cup V^R \cup V^W :$$

- *a set* V^{RI} *of nodes representing the* `READ`*-phase for inputs,*
- *a set* V^{WO} *of nodes representing the* `WRITE`*-phase for outputs,*
- *a set* V^R *of nodes representing the* `READ`*-phase for wires,*
- *a set* V^W *of nodes representing the* `WRITE`*-phase for wires.*

In addition to the nodes, G^P *contains*

- *a set of edges* $E \subseteq V \times V$, *representing interconnections between nodes,*
- *a set of system components* $C = \{c_1, \ldots, c_{n_C}\}$, *and*
- *an instantiation function* $I : V^I \to C$ *defining by* $I(v_i) = c_l$ *that* $v_i \in V^I$ *is an instance of system component* c_l.

The generation of the partitioning graph is illustrated in figure 4.20 and described by algorithm 1. First, the hierarchy of the system is flattened recursively by replacing the structural components by their netlists (line 11). The result is a netlist wiring all instances of behaviorally specified system components.

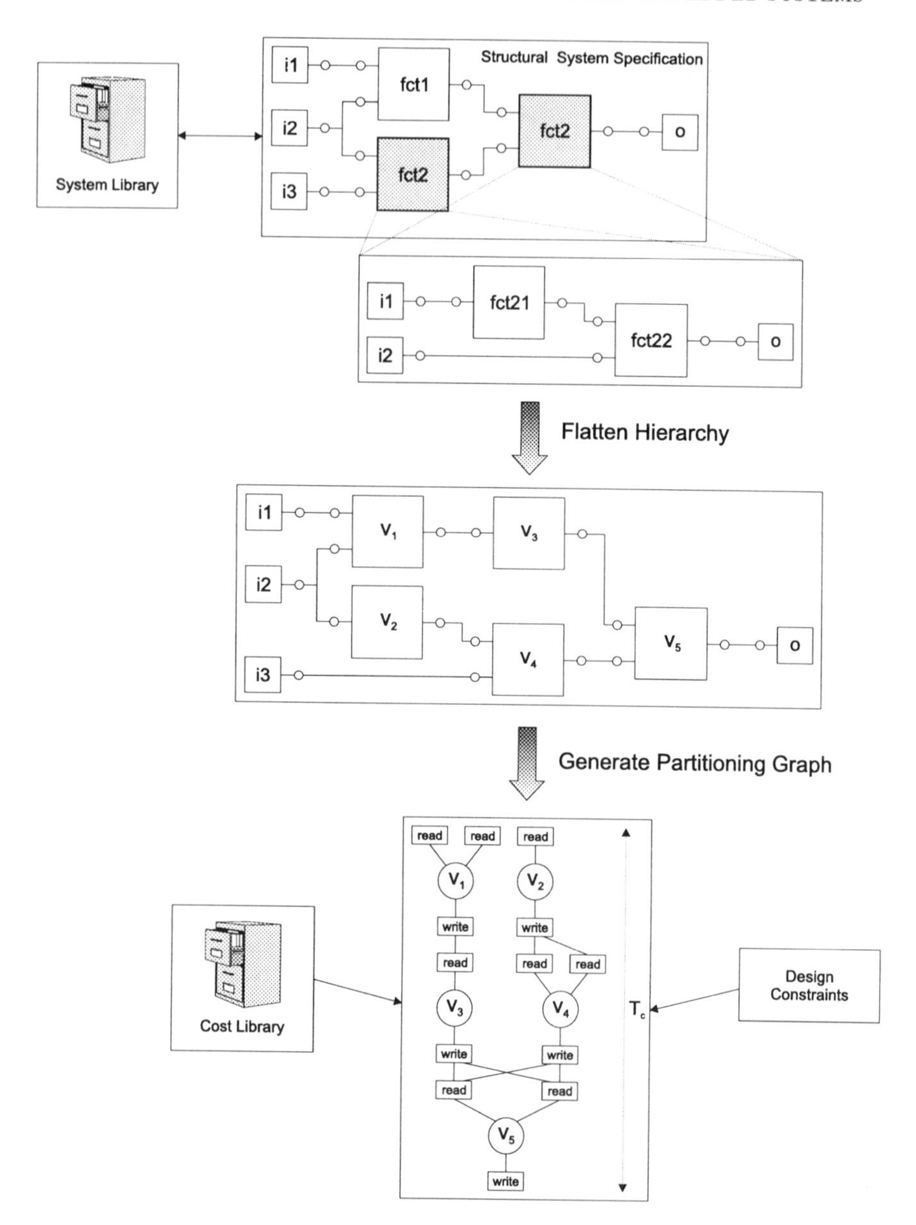

Figure 4.20. Generation of the partitioning graph

These components are collected in line 12. Then, the partitioning graph is created by adding a computation node for each instance of a component of the flattened system (lines 13 and 14). Communication nodes representing data transfers by READ- and WRITE-phases are inserted in lines 15 to 24. All these nodes are connected by edges corresponding to the signals of the flattened system specification (line 27). Furthermore, additional edges between WRITE- and READ-nodes are inserted (lines 28 to 36) to guarantee the *execution order* described in section 4.1.4. After the structure of the graph has been computed, computation nodes $v \in V^I$ are weighted with hardware and software costs. Communication nodes $v \in V^{RW}$ are weighted with additional interface costs stored in the cost library (line 37). Finally, user-defined design constraints are attached to the graph (line 38). Thus, the partitioning graph includes all information required for partitioning.

Algorithm 1: System2PGraph

```
(1)  algorithm System2PGraph;
(2)      input S : StructuralSystem;
(3)      output G^P = (V,E,C,I) : PartitioningGraph;
(4)  {
(5)      variable S'                 : StructuralSystem;
(6)      variable i, i_1, i_2        : Behavioral_Instance_of_a_System;
(7)      variable w                  : Wire_of_a_System;
(8)      variable L_read, L_write    : list of node;
(9)      variable v_i, v_1, v_2      : node;
(10)
(11)     S' = FlattenHierachy(S);
(12)     C = GetComponents(S');
(13)     ∀ instances i wired in S' do
(14)       V^I.insert(new computation_node(i));
(15)     ∀ wires w connected to an input signal in S' do
(16)       V^RI.insert(new READ-node(w));
(17)     ∀ wires w connected to an output signal in S' do
(18)       V^WO.insert(new WRITE-node(w));
(19)     ∀ wires between two instances i_1, i_2 in S' do
(20)     {
(21)       V^W.insert(new WRITE-node(i_1, i_2));
(22)       V^R.insert(new READ-node(i_1, i_2));
(23)     }
(24)     V^RW = V^RI ∪ V^WO ∪ V^R ∪ V^W;
(25)     V = V^I ∪ V^RW;
(26)     I=GetInstantiationFunction(C,V);
(27)     E=CreateEdges(S');
(28)     ∀ v_i ∈ V^I do
(29)     {
(30)       L_read = Get_Predecessors(G^P, v_i);
(31)       if | L_read |≥ 2 then
(32)         L_write = Get_Predecessing_Writing_Nodes(G^P, L_read);
(33)         ∀ v_1 ∈ L_write do
(34)           ∀ v_2 ∈ L_read do
(35)             G^P.InsertEdge(v_1, v_2);
(36)     }
(37)     Annotate_Costs(G^P, S');
(38)     Annotate_Constraints(G^P, S');
```

The following example illustrates the generation of the partitioning graph with the help of a 4-band equalizer. This 4-band equalizer will be the demonstrator example for most of the aspects in hardware/software partitioning which will be presented in the following sections.

Example 7:

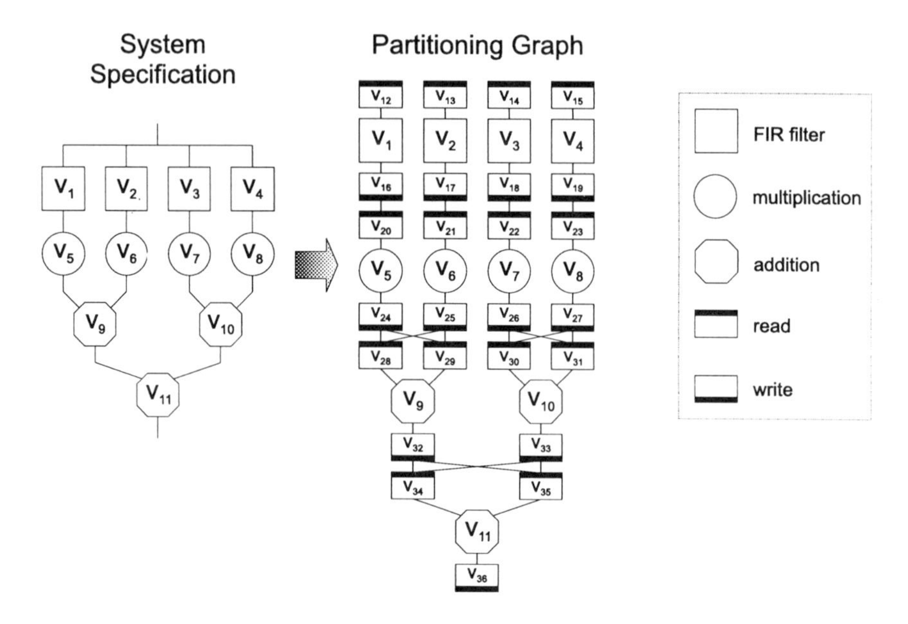

Figure 4.21. Partitioning graph

In figure 4.21, the structural system specification is depicted. It consists of 11 instances of three system components: an FIR-filter, a multiplication and an addition. This system specification is transformed into a partitioning graph by the following steps: First, for each instance $(v_1, \ldots, v_{11})$ a node is inserted. Then, for each access to an input signal nodes $(v_{12}, \ldots, v_{15})$ are added representing the READ-phases for these inputs. Furthermore v_{36} is added representing the WRITE-phase for the computed output signal. For each wire of the system specification a WRITE- and a READ-node are inserted $(v_{16}, \ldots, v_{35})$. Afterwards, all these nodes are connected by edges corresponding to the wiring of the system specification. Finally, additional edges are inserted $(v_{24}, v_{29}), (v_{25}, v_{28}), (v_{26}, v_{31}), (v_{27}, v_{30}), (v_{32}, v_{35}), (v_{33}, v_{34})$ to guarantee the execution order.

4.6 Formulation of the HW/SW Partitioning Problem

The hardware/software partitioning problem will be modelled in the following by an MILP model. Therefore, this section introduces a formulation of the hardware/software partitioning problem to simplify the description of the MILP model. The inputs for the partitioning problem include the system specification in form of a *partitioning graph* (with *costs* and *design constraints* attached to it) and the *target technology*. In this section, the terms necessary to describe the hardware/software partitioning problem will be defined formally. The partitioning graph has already been defined in section 4.5.

Target Technology. A target technology consists of processors, ASICs and memories which are connected by communication channels. A formal definition will be given:

Definition 2: *Target technology*

A **target technology** $\mathcal{T}$ *is defined as a tuple*

$$\mathcal{T} = (\mathcal{V}, \mathcal{E}), \qquad \mathcal{V} = \mathcal{H} \cup \mathcal{P} \cup \mathcal{M}, \qquad \mathcal{E} \subseteq \mathcal{PS}(\mathcal{V}) \setminus \Big\{ \{v\} \mid v \in \mathcal{V} \Big\}$$

containing target technology components and communication channels. The target technology components $\mathcal{V}$ are defined as a set of hardware components (ASICs) $\mathcal{H} = \{h_1, \ldots, h_{n_\mathcal{H}}\}$, processors $\mathcal{P} = \{p_1, \ldots, p_{n_\mathcal{P}}\}$ and memories $\mathcal{M} = \{m_1, \ldots, m_{n_\mathcal{M}}\}$. The processing units $\mathcal{PU} = \mathcal{P} \cup \mathcal{H}$ are defined as the union of processors and ASICs. The communication channels $\mathcal{E} = \{e_1, \ldots, e_{n_\mathcal{E}}\}$ connect target technology components (at least 2) where $\mathcal{PS}(\mathcal{V})$ represents the power set of $\mathcal{V}$.

A single-processor-single-ASIC target architecture is used to illustrate def. 2.

Example 8:

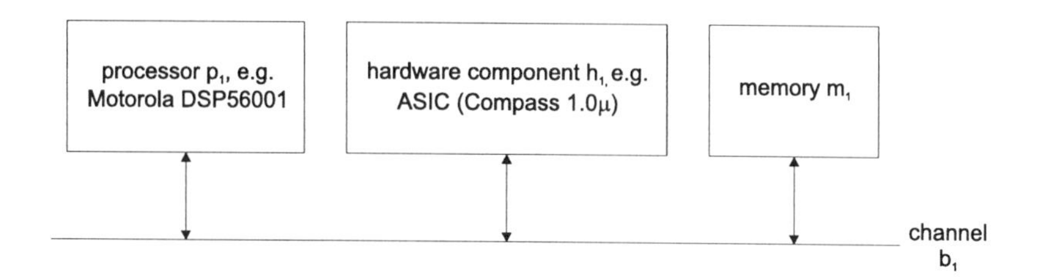

Figure 4.22. Target technology

In figure 4.22, an example of a target technology is given. It contains a processor $p_1 \in \mathcal{P}$, a hardware component $h_1 \in \mathcal{H}$, a memory $m_1 \in \mathcal{M}$ and a communication channel $b_1 \in \mathcal{E}$ connecting p_1, h_1 and m_1. Processor p_1 and ASIC h_1 represent the processing units of the target technology ($p_1, h_1 \in \mathcal{PU}$). The processor is a MOTOROLA DSP56001 and the ASIC should be manufactured by using a 1.0μ component library. The communication between processor and hardware component is implemented via *shared memory*.

Hardware and Software Implementation. Computation nodes of the partitioning graph may be implemented in hardware or in software. The main difference between implementing nodes v_{i_1}, v_{i_2} on a processor or on a hardware component is that v_{i_1} and v_{i_2} can not be executed concurrently on a processor (see section 4.1.1). On the software side, a system component c_l is implemented as a function on a processor. Each node v_i (representing an instance of c_l) which is mapped to the processor uses a corresponding function call for this function. On the hardware side however, two nodes v_{i_1} and v_{i_2} may be executed concurrently on a hardware component. Therefore, it is possible that v_{i_1} and v_{i_2} are mapped to different hardware instances of c_l. Definition 3 will introduce different implementation possibilities.

Definition 3: *Hardware and software implementations*

Let $G^P = (V, E, C, I)$ *be a partitioning graph.*
$\mathcal{T} = (\mathcal{V}, \mathcal{E})$ *be a target technology with* $\mathcal{V} = \mathcal{H} \cup \mathcal{P} \cup \mathcal{M}$ *(see def. 2).*
$p_k \in \mathcal{V}$ *be a processor and* $h_k \in \mathcal{V}$ *a hardware component.*

The set of possible **hardware implementations** $Impl^{hw}(c_l, h_k)$ *for a system component* c_l *is defined as:*

$$Impl^{hw}(c_l, h_k) = \left\{ h_{1,l,k} \ldots h_{N,l,k} \mid N = \left| \{v_i | I(v_i) = c_l\} \right| \right\}$$

where $h_{j,l,k}$ *represents the* j*-th* **hardware instance** *of* c_l *on* h_k.
The set of possible **software implementations** $Impl^{sw}(c_l, p_k)$ *for a system component* c_l *is defined as:*

$$Impl^{sw}(c_l, p_k) = \{p_{l,k}\}$$

where $p_{l,k}$ *represents a* **software function** *implementing* c_l *on* p_k.
The sets of possible hardware implementations $Impl^{hw}(G^P, \mathcal{H})$ *and software implementations* $Impl^{sw}(G^P, \mathcal{P})$ *for the partitioning graph* G^P *on* $\mathcal{T}$ *are then defined as:*

$$Impl^{hw}(G^P, \mathcal{H}) = \bigcup_{c_l \in C, h_k \in \mathcal{H}} Impl^{hw}(c_l, h_k)$$

$$Impl^{sw}(G^P, \mathcal{P}) = \bigcup_{c_l \in C, p_k \in \mathcal{P}} Impl^{sw}(c_l, p_k)$$

Example 9:

The possible hardware/software implementations for the 4-band equalizer are depicted in figure 4.23. Three functions may be implemented in software, one for an FIR-filter (`FIR`), one for multiplying (`MUL`) and one function for adding (`ADD`). On the hardware side, four hardware instances of an FIR-filter may be required ($\texttt{FIR}_1, \ldots, \texttt{FIR}_4$). In such a case, the highest performance can be obtained, because all four hardware instances are able to work concurrently. Finally, the hardware may contain a maximum of four hardware multipliers ($\texttt{MUL}_1, \ldots, \texttt{MUL}_4$) and three hardware adders ($\texttt{ADD}_1, \ldots, \texttt{ADD}_3$).

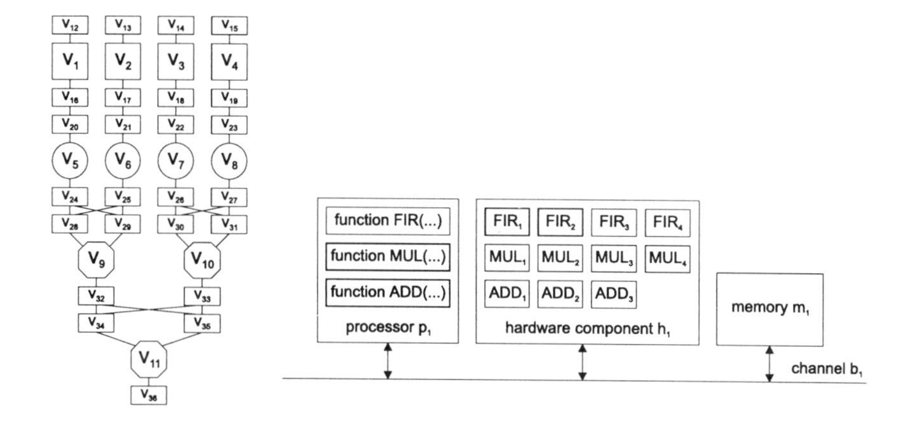

Figure 4.23. Hardware/software implementations

Cost Model. The nodes of the partitioning graph are weighted with additional costs as described in section 4.4. The following notations for the cost metrics are used in this book.

Definition 4: *Cost model*

Let $G^P = (V, E, C, I)$ *be a partitioning graph.*
$V = V^I \cup V^{RI} \cup V^{WO} \cup V^R \cup V^W$.
$c \in C$ *be a system component and* $v \in V$ *a node of* G^P.
$\mathcal{T} = (\mathcal{V}, \mathcal{E})$ *a target technology with* $\mathcal{V} = \mathcal{H} \cup \mathcal{P} \cup \mathcal{M}$ *(see def. 2).*
$p \in \mathcal{P}$ *be a processor,* $h \in \mathcal{H}$ *a hardware component,*
$b \in \mathcal{E}$ *a communication channel.*

The **cost metrics** *are defined as follows:*

$c^{dm}(c,p)$ *represents the software data memory required by* c *on* p,
$c^{pm}(c,p)$ *the software program memory required by* c *on* p,
$c^{ts}(c,p)$ *the software execution time required by* c *on* p,
$c^{a}(c,h)$ *the hardware area required by* c *on* h,
$c^{th}(c,h)$ *the hardware execution time required by* c *on* h,
$c^{dm}(v,p)$ *represents the software data memory required by* $v \in V^I$ *on* p,
$c^{pm}(v,p)$ *the software program memory required by* $v \in V^I$ *on* p,
$c^{ts}(v,p)$ *the software execution time required by* $v \in V^I$ *on* p,
$c^{a}(v,h)$ *the hardware area required by* $v \in V^I$ *on* h,
$c^{th}(v,h)$ *the hardware execution time required by* $v \in V^I$ *on* h,
$c^{tw}(v,b)$ *the writing communication time for* $v \in V^{WO} \cup V^W$ *using* b,
$c^{tr}(v,b)$ *the reading communication time for* $v \in V^{RI} \cup V^R$ *using* b.

The costs for different nodes $v_{i_1}, v_{i_2} \in V^I$, representing the same system component, are obviously equal. The following equations reflect this fact:

Let $G^P = (V, E, C, I)$ be a partitioning graph; $V = V^I \cup V^{RW}$ (see def. 1).
$\mathcal{T} = (\mathcal{V}, \mathcal{E})$ be a target technology with $\mathcal{V} = \mathcal{H} \cup \mathcal{P} \cup \mathcal{M}$ (see def. 2)
$p \in \mathcal{P}$ be a processor and $h \in \mathcal{H}$ a hardware component of $\mathcal{T}$.

$\forall v_{i_1}, v_{i_2} \in V^I$:

$$I(v_{i_1}) = I(v_{i_2}) \Rightarrow \forall p \in \mathcal{P}: \quad c^{ts}(v_{i_1}, p) = c^{ts}(v_{i_2}, p) \wedge c^{dm}(v_{i_1}, p) = c^{dm}(v_{i_2}, p) \wedge c^{pm}(v_{i_1}, p) = c^{pm}(v_{i_2}, p) \tag{4.3}$$

$$I(v_{i_1}) = I(v_{i_2}) \Rightarrow \forall h \in \mathcal{H}: \quad c^{th}(v_{i_1}, h) = c^{th}(v_{i_2}, h) \wedge c^{a}(v_{i_1}, h) = c^{a}(v_{i_2}, h) \tag{4.4}$$

After having defined the cost metrics for system components and nodes, the resource costs for each processing unit will be defined now. The resource costs of a processing unit t_k represent the sum of cost metrics required by the nodes mapped to t_k. For each of these resource costs maximum values can be defined. These are called resource constraints and represent, for example, the maximum number of CLBs of an FPGA or the amount of internal memory of a processor. Definition 5 will introduce resource costs and resource constraints.

Definition 5: *Resource costs and constraints*

Let $G^P = (V, E, C, I)$ *be a partitioning graph.*
$\mathcal{T} = (\mathcal{V}, \mathcal{E})$ *be a target technology with* $\mathcal{V} = \mathcal{H} \cup \mathcal{P} \cup \mathcal{M}$ *(see def. 2)*
$p \in \mathcal{P}$ *be a processor and* $h \in \mathcal{H}$ *a hardware component of* $\mathcal{T}$.

The **resource costs** *required for implementing* G^P *on processing units are defined as follows:*

$C^{dm}(p)$ *represents the software data memory required on* p,
$C^{pm}(p)$ *the software program memory required on* p,
$C^{a}(h)$ *the hardware area required on* h.

$MAX^{dm}(p)$, $MAX^{pm}(p)$ *and* $MAX^{a}(h)$ *represent the* **resource constraints** *according to the defined resource costs.*

A **design** represents the implementation of a partitioning graph on a target technology. The **design quality** can be expressed by evaluating the resource costs. The resulting costs are called **design costs**. They may also be constrained by **design constraints**. For example, the total execution time of a system has to fulfill a timing constraint to guarantee real-time conditions. In definition 6, design costs and design constraints are specified.

Definition 6: *Design costs and constraints*

Let $G^P = (V, E, C, I)$ *be a partitioning graph.*
$\mathcal{T} = (\mathcal{V}, \mathcal{E})$ *a target technology.*

The **design costs** *of a partitioning graph* G^P *are defined as follows:*

$C^{dm}(G^P, \mathcal{T})$ *represents the required software data memory,*
$C^{pm}(G^P, \mathcal{T})$ *the required software program memory,*
$C^{a}(G^P, \mathcal{T})$ *the hardware area and*
$C^{t}(G^P, \mathcal{T})$ *the total execution time.*

$MAX^{dm}(G^P, \mathcal{T}), MAX^{pm}(G^P, \mathcal{T}), MAX^{a}(G^P, \mathcal{T})$ *and* $MAX^{t}(G^P, \mathcal{T})$ *represent the* **design constraints** *according to the defined design costs.*

Using this complex cost model, the hardware/software partitioning problem can now be defined.

Hardware/Software Partitioning. The purpose of the hardware/software partitioning is to map computation nodes of the partitioning graph to processing units and communication nodes if communication is required to communication channels (see figure 4.1). To define the hardware/software partitioning problem, the notion *hardware/software mapping* has to be introduced (see definition 7).

Definition 7: *Hardware/software mapping*

Let $G^P = (V, E, C, I)$ *be a partitioning graph;* $V = V^I \cup V^{RW}$ *(def. 1).*
$\mathcal{T} = (\mathcal{V}, \mathcal{E})$ *be a target technology;* $\mathcal{V} = \mathcal{H} \cup \mathcal{P} \cup \mathcal{M}$ *(def. 2)*
$p_k \in \mathcal{P}$ *be a processor,* $h_k \in \mathcal{H}$ *a hardware component.*
$b_k \in \mathcal{E}$ *a communication channel.*
$Impl^{hw}(G^P, \mathcal{H})$ *be the set of possible hardware impl. of* G^P *on* $\mathcal{H}$.
$Impl^{sw}(G^P, \mathcal{P})$ *be the set of possible software impl. of* G^P *on* $\mathcal{P}$.

A **hardware/software mapping** $Map(G^P, \mathcal{T})$ *is defined as a tuple*

$$Map(G^P, \mathcal{T}) = (Map_{\mathcal{V}}, Map_{\mathcal{E}})$$

consisting of two sets $Map_{\mathcal{V}}$ *and* $Map_{\mathcal{E}}$ *which are defined as follows:*

$Map_{\mathcal{V}} \subseteq Impl^{hw}(G^P, \mathcal{H}) \cup Impl^{sw}(G^P, \mathcal{P})$ *represents the set of required hardware and software implementations to implement all computation nodes of the partitioning graph on processors and ASICs.*

$Map_{\mathcal{E}} \subseteq \mathcal{E}$ *represents the set of required communication channels to implement all necessary communication nodes.*

A hardware/software mapping of G^P to $\mathcal{T}$ can be described by two **mapping functions***:*

$$
\begin{aligned}
mv: \quad & V^I & \rightarrow Map_{\mathcal{V}} \\
me: \quad & V^{RW} & \rightarrow Map_{\mathcal{E}}
\end{aligned}
$$

defined by

$$
mv(v_i) = \begin{cases} p_{l,k} \in Impl^{sw}(c_l, p_k), & \text{if } v_i \text{ is implemented by function } p_{l,k} \text{ on } p_k \text{ calculating } c_l = I(v_i), \\ h_{j,l,k} \in Impl^{hw}(c_l, h_k), & \text{if } v_i \text{ is implemented by the } j\text{-th hardware instance } h_{j,l,k} \text{ of } c_l = I(v_i) \text{ on } h_k. \end{cases}
$$

$$
me(v_i) = \begin{cases} b_k \in \mathcal{E}, & \text{if } v_i \text{ is required to implement an interface on channel } b_k, \\ \emptyset, & \text{if no interface is required for } v_i. \end{cases}
$$

The goal of partitioning algorithms presented in this book is to minimize a user-defined objective function considering the design costs while meeting all performance requirements. The design costs are calculated using the given cost model. With the help of the introduced notations, the hardware/software partitioning problem can be formulated by definition 8:

Definition 8: *Hardware/software partitioning problem*

Let $G^P = (V, E, C, I)$ *be a partitioning graph;* $V = V^I \cup V^{RW}$ *(def. 1).*
$\mathcal{T} = (\mathcal{V}, \mathcal{E})$ *be a target technology;* $\mathcal{V} = \mathcal{H} \cup \mathcal{P} \cup \mathcal{M}$ *(def. 2)*
$p_k \in \mathcal{P}$ *be a processor,* $h_k \in \mathcal{H}$ *a hardware component.*
$b_k \in \mathcal{E}$ *a communication channel.*
$Map(G^P, \mathcal{T}) = (Map_{\mathcal{V}}, Map_{\mathcal{E}})$ *be a mapping of* G^P *onto* T *(def. 7).*

The **hardware/software partitioning problem** *is defined as the problem of finding the best hardware/software mapping of* G^P *to* $\mathcal{T}$ *which minimizes the objective function while meeting all resource and design constraints.*

4.7 Optimization

Real-life problems can very often be modelled as optimization problems including three basic ingredients:

1. a set of *variables* representing the unbounded values of the problem that should be optimized,

2. an *objective function* which should be minimized or maximized, and
3. a set of *constraints* that describe the relation between the variables, restricting the solution space.

Two different approaches for solving optimization problems,

- integer programming (IP) and
- genetic algorithms (GA),

will be described in the following. Both approaches have been used for solving the hardware/software partitioning problem in this book.

4.7.1 *Linear and Integer Linear Programming*

The basic idea of *linear programming (LP)* is to formulate a precise mathematical model for the given problem, containing an objective function, a set of variables and (in)equations. Linear programming, coming from *operations research*, has gained great importance. The word "programming" is used here in the sense of "modelling the problem", because no algorithm is implemented for solving the problem, but a mathematical formulation is specified. The way of solving this model is not described in the linear programming model. The *linear programming problem* is defined as follows:

Definition 9: *Linear programming problem*

Given a set of real numbers

$$b_i, c_j, a_{i,j} \in \mathbb{R} \;, \forall i \in \{1 \ldots m\} \subseteq \mathbb{N}, \forall j \in \{1 \ldots n\} \subseteq \mathbb{N}$$

Minimize the objective function:

$$Z(x_1, \ldots, x_n) = \sum_{j=1}^{n} c_j * x_j$$

while fulfilling the constraints

$$\forall i \in \{1 \ldots m\} : \sum_{j=1}^{n} a_{i,j} * x_j \leq b_i$$

where $x_1, \ldots, x_n \in \mathbb{R}$ *are the solution variables.*

Such linear programming problems, containing m constraints and n variables, can be solved by the so-called *simplex algorithm* [Neum75], described in more detail in [Wege93]. The simplex algorithm has an exponential worst-case computation time. Khachian proved in 1979 that the linear programming problem can be solved in polynomial time using his *ellipsoid method* [Khac79]. Another polynomial algorithm, called *interior-point method* [Karm84], has been developed by Karmarkar in 1984. Although the simplex algorithm has an exponential worst-case computation time, it is used in practice. The reason is that the ellipsoid method cannot compete with the simplex algorithm concerning computation time and the interior-point method lacks robustness.

The *integer linear programming problem* restricts the linear programming problem by the additional requirement of x_j having integer instead of real values.

Definition 10: *Integer programming problem*

Given a set of real numbers

$$b_i, c_j, a_{i,j} \in \mathbb{R}\ , \forall i \in \{1 \ldots m\} \subseteq \mathbb{N}, \forall j \in \{1 \ldots n\} \subseteq \mathbb{N}$$

Minimize the objective function:

$$Z(x_1, \ldots, x_n) = \sum_{j=1}^{n} c_j * x_j$$

while fulfilling the constraints

$$\forall i \in \{1 \ldots m\} : \sum_{j=1}^{n} a_{i,j} * x_j \leq b_i$$

where $x_1, \ldots, x_n \in \mathbb{Z}$ *are the solution variables.*

At first sight, integer and linear programs are very similar. However, it has been shown that the integer programming problem is NP-hard [GaJo79]. This is proven by the fact that the NP-hard knapsack problem is a special case of the IP-problem (see [Wege93]). The intuitive reason for this difference in complexity is that linear programming represents a *contiguous* and integer programming a *discrete problem*. There are a variety of algorithms solving the IP-problem. Some approaches are based on the simplex-algorithm including *branch-and-bound methods* [Mitr73] and the *cutting-plane method* [Gomo60]. Other approaches use BDDs[11] [LPV94] for solving IP problems. A detailed overview of algorithms for solving LP- and IP-problems is given in [Nemh88].

Two other related problems should be presented:

[11] BDD: Binary Decision Diagram

1. If the variables $x_1, \ldots, x_n$ are restricted to binary values $x_1, \ldots, x_n \in \{0, 1\}$, the problem is called the *binary programming problem (0-1-IP)*.
2. If only a subset of the solution variables is restricted to integer values, the resulting problem is called a *mixed integer linear programming problem (MILP)*.

The use of mixed integer linear programming for solving the hardware/software partitioning problem will be described in this book. The advantages of using integer programming can be summarized as follows:

- modelling a problem is much easier than specifying an algorithm solving the problem,
- specifying an IP model helps to understand the problem,
- all constraints are taken into account simultaneously,
- optimal results for the objective function are computed,
- a variety of tools is available for solving the specified problem.

The disadvantage of IP is the high computation time for larger problems.

4.7.2 Evolutionary Algorithms

Another approach for solving optimization problems is based on imitating natural optimization processes. These algorithms are called *evolutionary algorithms (EA)*. These algorithms are probabilistic and simulate the evolution process by generating a number of random solutions. Some of these solutions are selected using an evaluation function and variated randomly afterwards. Evolutionary algorithms iteratively improve the solution, applying small modifications to good solutions to obtain better solutions.

4.7.2.1 The Evolutionary Algorithm. The basic evolutionary algorithm is depicted in figure 4.24.

In a first step, a set of solutions (*individuals*) is generated. This set of individuals is called a *population*. The *initial population* P_0 represents the first *generation* during the evolutionary process. Then, the evolutionary algorithm iteratively computes new populations until the maximal number N of generations has been reached or a *stopping criterion* is fulfilled. During each generation i all individuals of the population are examined. In some cases, the individuals are manipulated by a *repair mechanism* to obtain valid solutions. Then, all individuals are evaluated by a *fitness function* F. During the *selection* phase, the best individuals $P_i^{'} \subset P_i$ from population P_i are selected, considering their *fitness*. During a *variation* phase, a new population P_{i+1} is computed, varying the individuals of $P_i^{'}$ by evolutionary operators. Finally, after all iter-

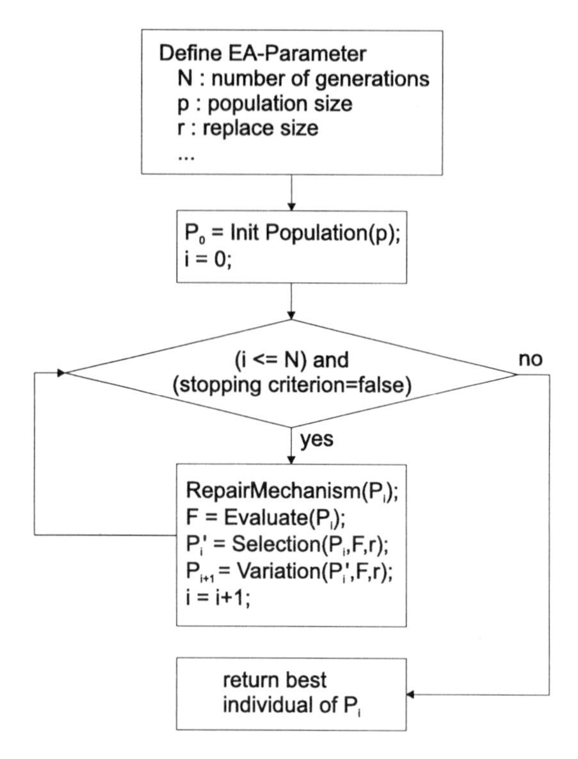

Figure 4.24. General evolutionary algorithm

ations have been executed or the stopping criterion has been fulfilled, the best individual represents the final solution.

4.7.2.2 Evolutionary Operators. *Evolutionary operators* vary a given solution to produce new modified solutions. These operators can be found in all approaches to evolutionary algorithms, but in this approach, genetic algorithms will be used. Therefore, we will introduce some special vocabulary used in the GA domain to describe genetic operators.

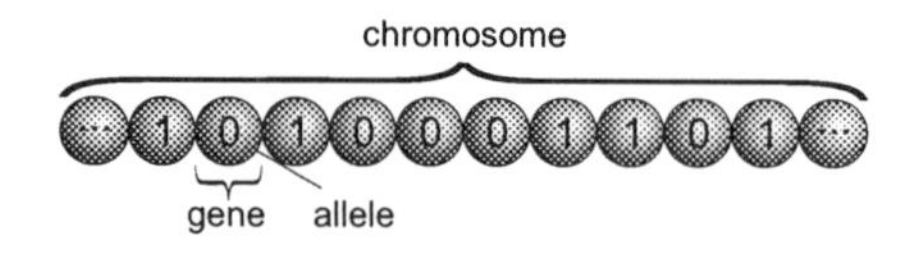

Figure 4.25. Genetic algorithm encoding

In figure 4.25, the way of encoding solutions using genetic algorithms is depicted. The solution *string* is represented as a *chromosome* consisting of a set of *genes*. The value which is encoded by the gene is called *allele*. In most

Algo.	Coding	Selection	Recombination	Mutation
ES	real numbers	deterministic	yes	yes
EP	real numbers	stochastic	no	yes
GA	discrete numbers	stochastic	mainly	yes

Table 4.3. Overview of evolutionary algorithms

approaches these alleles are binary values, but in some other approaches they represent discrete values in general.

The variation step includes the execution of two basic evolutionary operators as depicted in figure 4.26.

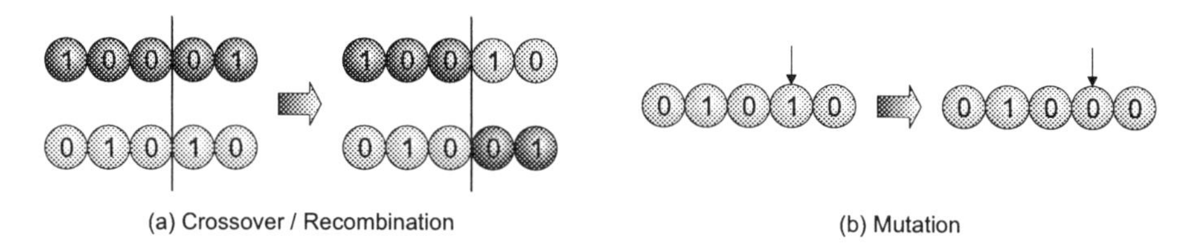

Figure 4.26. Crossover and mutation

1. The *crossover* or *recombination* operator splits two chromosomes and recombines them as depicted in figure 4.26(a). This operator is called *single point crossover*.
2. The *mutation* operator changes only one gene of the chromosome as illustrated in figure 4.26(b).

4.7.2.3 Comparison of Evolutionary Algorithms. There are three major approaches to evolutionary algorithms. They differ in some details including the way of encoding the solutions, the selection mechanism and in applying evolutionary operators. In table 4.3, a comparison of the approaches is given.

Evolutionary strategies (ES) have been introduced by Schwefel [Schw75] using real numbers to encode the solutions. The selection scheme is deterministic, selecting only the best solutions. Both, mutation and recombination operators are used during the optimization process. *Evolutionary programming (EP)* is very similar to evolutionary strategy. The coding is also done using real numbers, but mutation is the only variation operator. In contrast to evolutionary strategies, the selection is based on a stochastic method, enabling bad solutions to be inherited with a certain probability. *Genetic algorithms (GA)* [Holl75, Gold89] have been developed by Holland. They use discrete and in most cases binary values for encoding the solutions. The recombination operator is the most important one, but mutation is possible. The selection operator works stochastically.

Genetic algorithms have been used for solving the hardware/software partitioning problem and for performing design space exploration in COOL, because of their following advantages:

- the computation time is predictable,
- GA's can also be applied to *non-linear optimization problems*, in contrast to IP which can only model *linear optimization problems*,
- GA's can also be applied to NP-hard or NP-complete problems where exact methods like IP fail due to computation time,
- GA's are able to overcome *local extrema* by mutation and crossover,
- a set of different solutions is generated during the optimization process supporting *design space exploration*.

The disadvantage of genetic algorithms and all other heuristic algorithms is the loss of optimality, but in practice it has been shown that the results are often of high quality.

4.8 The MILP-Model

Many optimization problems can be solved optimally by using mixed integer linear programming (MILP). In the following, an MILP model will be presented, allowing to solve the hardware/software partitioning problem presented in sections 4.1 and 4.3.1. To simplify the description of the MILP model, the following notations are introduced:

Definition 11: *Sets of nodes*

Let $G^P = (V, E, C, I)$ *be a partitioning graph;* $V = V^I \cup V^{RW}$ *(def. 1).*

$$
\begin{array}{lll}
pred\_nodes(v \in V) & = \{w \in V & \mid \exists p : p = (w, \ldots, v)\} \\
succ\_nodes(v \in V) & = \{w \in V & \mid \exists p : p = (v, \ldots, w)\} \\
instances\_of(c \in C) & = \{w \in V^I & \mid I(w) = c\} \\
share\_nodes(v \in V) & = \{w \in V^I & \mid w \neq v \wedge I(v) = I(w)\} \\
schedule\_nodes(v \in V^I) & = \{w \in V^I & \mid w \neq v \wedge w \notin \{pred\_nodes(v) \cup succ\_nodes(v)\}\} \\
schedule\_nodes(v \in V^{RW}) & = \{w \in V^{RW} & \mid w \neq v \wedge w \notin \{pred\_nodes(v) \cup succ\_nodes(v)\}\} \\
path\_nodes(v_1, v_2 \in V) & = \{w \in V & \mid w \in succ\_nodes(v_1) \wedge w \in pred\_nodes(v_2)\} \\
dominator\_nodes(v \in V) & = \{w \in V & \mid \forall p : p = (s, \ldots, v) \wedge pred\_nodes(s) = \emptyset : w \in p\} \\
read\_nodes(v \in V^I) & = \{w \in V^I & \mid \forall e = (w, v) \in E \wedge w \in V^R \cup V^{RI}\} \\
write\_nodes(v \in V^I) & = \{w \in V^I & \mid \forall e = (v, w) \in E \wedge w \in V^W \cup V^{WO}\} \\
uses\_data\_of(v \in V^I) & = \{w \in V^I & \mid \exists p : p = (w, v_i, v_j, v) \wedge v_i \in V^W \wedge v_j \in V^R\}
\end{array}
$$

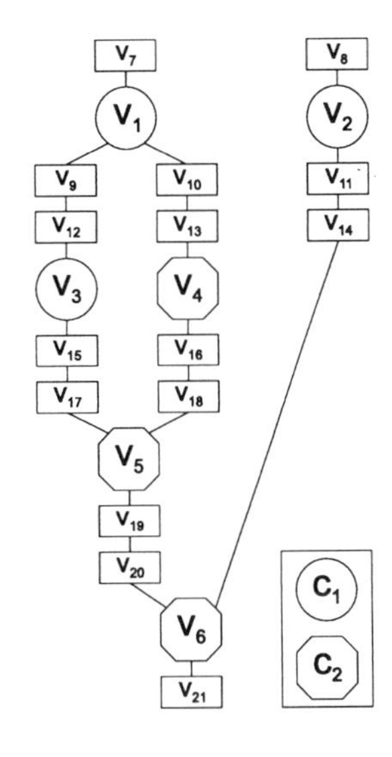

Figure 4.27. Node sets

In figure 4.27, an example is given to explain these sets informally, because they are essential for understanding the MILP model. A set *pred_nodes*(v) contains all predecessor nodes of v, e.g. *pred_nodes*$(v_3) = \{v_7, v_1, v_9, v_{12}\}$. According to this, *succ_nodes*(v) contains the successors of a node, e.g. *succ_nodes*$(v_5) = \{v_{19}, v_{20}, v_6, v_{21}\}$. The partitioning graph $G = (V, C, E, I)$ contains two different system components: $c_1, c_2 \in C$. The relation between system components $c \in C$ and nodes $v \in V^I$ is defined by the instantiation function I (see def. 1). The set *instances_of*(c_l) is used to collect all nodes representing instances of the same system component c_l. Therefore, *instances_of*$(c_1) = \{v_1, v_2, v_3\}$ and *instances_of*$(c_2) = \{v_4, v_5, v_6\}$. Instances of same system components c_l may share hardware resources. For this purpose, a set *share_nodes*(v) contains all nodes which may share resources with v, e.g. *share_nodes*$(v_1) = \{v_2, v_3\}$. Set *schedule_nodes*(v) for computation nodes contains all computation nodes which may be executed in parallel to v, e.g. *schedule_nodes*$(v_2) = \{v_1, v_3, v_4, v_5\}$. Set *schedule_nodes*$(v)$ for communication nodes is defined corresponding to the set for computation nodes. Therefore, *schedule_nodes*$(v_{10}) = \{v_8, v_{11}, v_{14}\}$. All computation nodes between two nodes are stored in *path_nodes*, e.g. *path_nodes*$(v_1, v_6) = \{v_3, v_4, v_5\}$. If all paths starting from an input node s to a node v contain the same node w, then w is called a dominator node of v. For each computation node v *dominator_nodes*(v) contains all dominator nodes of v, e.g. *dominator_nodes*$(v_5) = \{v_1\}$. A set *read_nodes*(v) contains all `READ`-nodes for computation node v and *write_nodes*(v) contains all `WRITE`-nodes of v, e.g. *read_nodes*$(v_5) = \{v_{17}, v_{18}\}$ and *write_nodes*$(v_5) = \{v_{19}\}$. Finally, the set *uses_data_of*(v) determines for each computation node v all other computation nodes w writing data which is read afterwards by v, e.g. *uses_data_of*$(v_5) = \{v_3, v_4\}$.

REMARK:. In the following description of the MILP problem, variables are used with indices. To simplify the description, always the same indices are used for the same purpose:

- $i \in \{1, \ldots, n_V\}$ is used to index nodes $v_i \in V$,
- $l \in \{1, \ldots, n_C\}$ represents the index for system components $c_l \in \mathcal{C}$,
- $k \in \{1, \ldots, n_\mathcal{V}\}$ is used as an index for target technology components, either
  - $k \in KH = \{1, \ldots, n_\mathcal{H}\}$ for hardware components $h_k \in \mathcal{H}$, or

- $k \in KP = \{1, \ldots, n_{\mathcal{P}}\}$ for processors $p_k \in \mathcal{P}$, or

- $k \in KB = \{1, \ldots, n_{\mathcal{E}}\}$ for communication channel $b_k \in \mathcal{E}$,

- $j \in \{1, \ldots, n_V\}$ is the index for the j-th hardware instance $h_{j,l,k}$ of a system component c_l mapped to hardware component h_k.

Using these indices, the notations for the cost metrics, the resource / design costs and constraints will be defined. All of them will be used to formulate the MILP model in the following. Definition 12 introduces all necessary notations.

Definition 12: *Variables for costs and constraints*

Let $G^P = (V, E, C, I)$ *be a partitioning graph;* $V = V^I \cup V^{RW}$ *(def. 1).*
$\mathcal{T} = (\mathcal{V}, \mathcal{E})$ *a target technology.*
$c_l \in C$ *be a system component and* $v_i \in V$ *a node of* G^P*.*
$p_k \in \mathcal{V}$ *be a processor,* $h_k \in \mathcal{V}$ *a hardware component.*
$b_k \in \mathcal{E}$ *a communication channel.*

$c_{i,k}^{ts}, c_{i,k}^{dm}, c_{i,k}^{pm}$	*cost metrics* $c^{ts}(v_i, p_k)$, $c^{dm}(v_i, p_k)$, $c^{pm}(v_i, p_k)$ *for* $v_i \in V^I$,
$c_{i,k}^{th}, c_{i,k}^{a}$	*cost metrics* $c^{th}(v_i, h_k)$, $c^{a}(v_i, h_k)$ *for* $v_i \in V^I$,
$c_{i,k}^{tw}, c_{i,k}^{tr}$	*cost metrics* $c^{tw}(v_i, b_k)$, $c^{tr}(v_i, b_k)$ *for* $v_i \in V^{RW}$,
C_k^{dm}	*resource cost* $C^{dm}(p_k)$,
C_k^{pm}	*resource cost* $C^{pm}(p_k)$,
C_k^{a}	*resource cost* $C^{a}(h_k)$,
MAX_k^{dm}	*resource constraint* $MAX^{dm}(p_k)$,
MAX_k^{pm}	*resource constraint* $MAX^{pm}(p_k)$,
MAX_k^{a}	*resource constraint* $MAX^{a}(h_k)$,
C^t	*design metric* $C^t(G^P, \mathcal{T})$,
C^{dm}	*design metric* $C^{dm}(G^P, \mathcal{T})$,
C^{pm}	*design metric* $C^{pm}(G^P, \mathcal{T})$,
C^{a}	*design metric* $C^{a}(G^P, \mathcal{T})$,
MAX^t	*design constraint* $MAX^t(G^P, \mathcal{T})$,
MAX^a	*design constraint* $MAX^a(G^P, \mathcal{T})$,
MAX^{dm}	*design constraint* $MAX^{dm}(G^P, \mathcal{T})$,
MAX^{pm}	*design constraint* $MAX^{pm}(G^P, \mathcal{T})$,
T_i^S	*starting time of node* v_i,
T_i^D	*execution time of node* v_i,
T_i^E	*ending time of node* v_i.

Some of these variables have discrete value ranges but most of them can take real numbers.

4.8.1 *The Decision Variables*

The MILP model uses decision variables for defining mapping, scheduling, sharing and interfacing constraints. In contrast to the variables for costs and constraints, which are represented by real numbers, the decision variables are represented by discrete, mostly binary, numbers. All decision variables used for the MILP problem are defined in definition 13.

Definition 13: *Decision Variables*

$$X_{i,k} = \begin{cases} 1 & : \textit{ } v_i \textit{ is mapped to hardware component } h_k, \\ 0 & : \textit{ otherwise.} \end{cases}$$

$$x_{i,j,k} = \begin{cases} 1 & : \textit{ } v_i \textit{ is mapped to the } j\textit{-th hardware instance of } \\ & \quad c = I(v_i) \textit{ on hardware component } h_k, \\ 0 & : \textit{ otherwise.} \end{cases}$$

$$Y_{i,k} = \begin{cases} 1 & : \textit{ } v_i \textit{ is mapped to processor } p_k, \\ 0 & : \textit{ otherwise.} \end{cases}$$

$$Z_i = \begin{cases} 1 & : \textit{ an interface is required for } v_i, \\ 0 & : \textit{ otherwise.} \end{cases}$$

$$z_{i,k} = \begin{cases} 1 & : \textit{ communication represented by } v_i \textit{ is implemented} \\ & \quad \textit{on communication channel } b_k, \\ 0 & : \textit{ otherwise.} \end{cases}$$

$$nx_{j,l,k} = \begin{cases} 1 & : \textit{ at least 1 instance of } c_l \textit{ is mapped to the } j\textit{-th} \\ & \quad \textit{hardware instance of } c_l \textit{ on } h_k, \\ 0 & : \textit{ otherwise.} \end{cases}$$

$$NY_{l,k} = \begin{cases} 1 & : \textit{ at least 1 instance of } c_l \textit{ is mapped to processor } p_k, \\ 0 & : \textit{ otherwise.} \end{cases}$$

$$NX_{l,k} \; : \; \textit{number of hardware instances of } c_l \textit{ implemented on } h_k.$$

$$b_{i_1,i_2} = \begin{cases} 1 & : \textit{ } v_{i_1} \textit{ ends before } v_{i_2} \textit{ starts,} \\ 0 & : \textit{ otherwise.} \end{cases}$$

The decision variables $X_{i,k}$ and $Y_{i,k}$ are used to model the mapping aspects for the computation nodes $v_i \in V^I$. Variable $x_{i,j,k}$ is used for hardware sharing, Z_i and $z_{i,k}$ model interfacing aspects for the communication nodes $v_i \in V^{RW}$. With the help of variables $nx_{j,l,k}$, $NX_{l,k}$ and $NY_{l,k}$, the resource costs can be accumulated. All b-variables are required to formulate the scheduling constraints between two nodes. The following example illustrates the application of these decision variables.

Example 10:

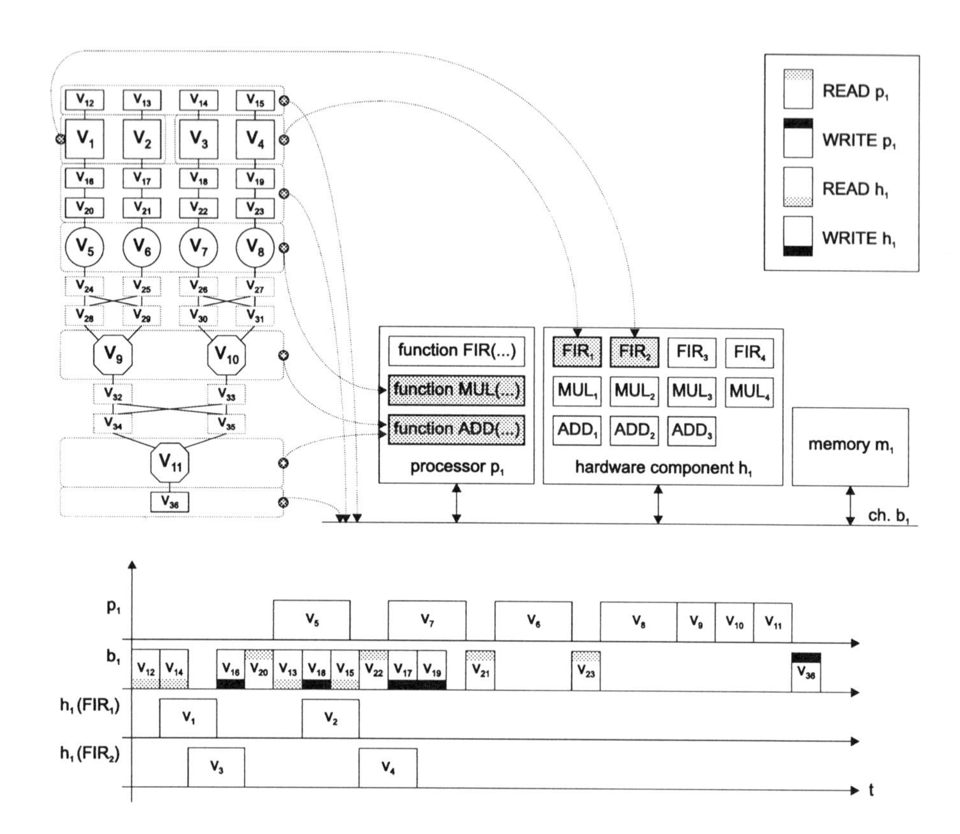

Figure 4.28. Usage of decision variables

In figure 4.28, a partitioning graph of the 4-band equalizer is depicted. Nodes $v_1, \ldots, v_4$ are mapped to hardware component h_1 ($X_{1,1} = \cdots = X_{4,1} = 1$). All other nodes $v_5, \ldots, v_{11}$ are mapped to processor p_1 ($Y_{5,1} = \cdots = Y_{11,1} = 1$). This mapping forces channel accesses for reading inputs ($Z_{12} = \cdots = Z_{15} = 1$), writing outputs ($Z_{36} = 1$) and exchanging data between hardware component h_1 and processor p_1 ($Z_{16} = \cdots = Z_{23} = 1$). Only one communication channel b_1 is available. Therefore, $z_{12,1} = \cdots = z_{23,1} = z_{36,1} = 1$. All other nodes ($v_{24}, \ldots, v_{35}$) are not necessary, because no communication is required ($Z_{24} = \cdots = Z_{35} = 0$). To reduce the amount of hardware area, v_1 and v_2 share the same hardware instance of an FIR-filter on h_1 ($x_{1,1,1} = x_{2,1,1} = 1$). Nodes v_3 and v_4 share the second one ($x_{3,2,1} = x_{4,2,1} = 1$). The nx-variables have the following values: $nx_{1,1,1} = nx_{2,1,1} = 1$ and $nx_{3,1,1} = nx_{4,1,1} = 0$, because only the first two hardware instances of four possible FIR-filters (system component c_1) are required on h_1 ($NX_{1,1} = 2$). All multiplications (c_2) and additions (c_3) are implemented on processor p_1. Therefore, one function is required for multiplying ($NY_{2,1} = 1$) and another function is required for adding ($NY_{3,1} = 1$). The timing diagram shows a possible schedule for this partitioning. In the depicted case the scheduling variable $b_{5,6}$ is 1, because v_5 is executed before v_6 on p_1.

4.8.2 *The Objective Function*

All optimization problems have a special *objective function* defining the goal of the optimization process. The general *objective function* for the hardware/software partitioning approach presented in this book is defined in equation 4.5.

$$minimze \; \{ \; f_1 * C^t + f_2 * C^{pm} + f_3 * C^{dm} + f_4 * C^a \; \} \qquad (4.5)$$

If, for example, the overall goal is to minimize hardware area under timing constraints, then $f_4 \gg f_1, f_2, f_3$. But other optimization goals can also be specified with this general objective function by changing the scaling factors f_1, f_2, f_3, f_4.

4.8.3 *The Constraints*

The following constraints have to be fulfilled:

1. **Mapping Constraints**: Each computation node $v_i \in V^I$ is executed exactly on one processing unit t_k, a processor or a hardware component (eq. 4.8).

$$\begin{array}{rrcll} \forall v_i \in V^I : \forall k \in KH : & X_{i,k} & \leq & 1 & (4.6) \\ \forall v_i \in V^I : \forall k \in KP : & Y_{i,k} & \leq & 1 & (4.7) \\ \forall v_i \in V^I : & \sum\limits_{k \in KH} X_{i,k} + \sum\limits_{k \in KP} Y_{i,k} & = & 1 & (4.8) \end{array}$$

In addition, it must be guaranteed that two nodes exchanging data are mapped to processing units which are connected by a communication channel. In figure 4.29 an example is given where processors p_1 and p_2 cannot communicate directly with each other.

In such a case, the mapping possibilities have to be eliminated (eq. 4.9-4.11).

$$\begin{array}{rrcll} \multicolumn{5}{l}{\forall v_{i_1}, v_{i_2} \in V^I : v_{i_2} \in uses\_data\_of(v_{i_1}):} \\ \forall k_1, k_2 \in KH : \nexists b \in \mathcal{E} : t_{k_1}, t_{k_2} \in b : & X_{i_1,k_1} + X_{i_2,k_2} & \leq & 1 & (4.9) \\ \forall k_1 \in KH, \forall k_2 \in KP : \nexists b \in \mathcal{E} : t_{k_1}, t_{k_2} \in b : & X_{i_1,k_1} + Y_{i_2,k_2} & \leq & 1 & (4.10) \\ \forall k_1, k_2 \in KP : \nexists b \in \mathcal{E} : t_{k_1}, t_{k_2} \in b : & Y_{i_1,k_1} + Y_{i_2,k_2} & \leq & 1 & (4.11) \end{array}$$

If a system component c_l has been implemented on a processor p_k, then it is not necessary to implement it more than once (eq. 4.12), because it can be implemented as <u>one</u> function and several function calls (see definition

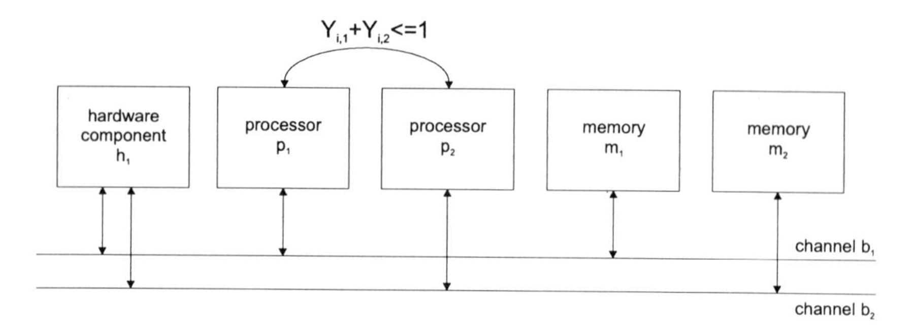

Figure 4.29. Constraints caused by target architecture restrictions

3). Therefore, the number $NY_{l,k}$ is calculated by equations 4.13 and 4.14. In contrast to the binary variable $NY_{l,k}$, the number of hardware instances $NX_{l,k}$ of a system component c_l implemented on a hardware component h_k may be greater than one. If no hardware sharing is considered, then $NX_{l,k}$ is equal to the sum of nodes, representing instances of c_l that have been mapped to h_k (eq. 4.15).

$\forall l \in L:$

$$\forall k \in KP: \quad NY_{l,k} \leq 1 \tag{4.12}$$

$$\forall v_i \in V^I : I(v_i) = c_l, \forall k \in KP: \quad NY_{l,k} \geq Y_{i,k} \tag{4.13}$$

$$\forall k \in KP: \quad NY_{l,k} \leq \sum_{\forall v_i \in V^I : I(v_i) = c_l} Y_{i,k} \tag{4.14}$$

$$\forall k \in KH: \quad NX_{l,k} = \sum_{\forall v_i \in V^I : I(v_i) = c_l} X_{i,k} \tag{4.15}$$

2. **Resource Constraints**: The hardware area C_k^a (eq. 4.16) used on a hardware component h_k is calculated by accumulating the costs for all hardware instances of system components implemented on h_k. The amount of required memory (eq. 4.17, 4.18) on a processor p_k is calculated by summing up the costs for implementing these system components as functions. The resource costs may not violate their resource constraints.

$$\forall k \in KH: \quad C_k^a = \sum_{l \in L} NX_{l,k} * c_{l,k}^a \leq MAX_k^a \tag{4.16}$$

$$\forall k \in KP: \quad C_k^{dm} = \sum_{l \in L} NY_{l,k} * c_{l,k}^{dm} \leq MAX_k^{dm} \tag{4.17}$$

$$\forall k \in KP: \quad C_k^{pm} = \sum_{l \in L} NY_{l,k} * c_{l,k}^{pm} \leq MAX_k^{pm} \tag{4.18}$$

3. **Design Constraints**: The design costs for the complete system are calculated by accumulating the resource costs required by the processing units of the target technology (eq. 4.19-4.21). In addition, these design costs may not exceed their given design constraints.

$$C^a = \sum_{k \in KH} C_k^a \leq MAX^a \tag{4.19}$$

$$C^{dm} = \sum_{k \in KP} C_k^{dm} \leq MAX^{dm} \tag{4.20}$$

$$C^{pm} = \sum_{k \in KP} C_k^{pm} \leq MAX^{pm} \tag{4.21}$$

4. **Timing Constraints**: The timing costs cannot be calculated by accumulating the execution time of nodes, because two nodes v_1, v_2 can concurrently be executed which reduces the total execution time. To determine the starting time and ending time for each node, scheduling has to be performed. The execution time T_i^D of node v_i is either a hardware or a software execution time for $v_i \in V^I$ (eq. 4.22), a reading execution time for $v_i \in V^{RI} \cup V^R$ (eq. 4.23) or a writing execution time for $v_i \in V^{WO} \cup V^W$ (eq. 4.24). The ending time T_i^E (eq. 4.25) of v_i is the sum of starting time T_i^S and execution time T_i^D. The system execution time C^t (eq. 4.26) is the maximum of the ending times of all nodes v_i and may not violate the global design timing constraint. Data dependencies (eq. 4.27) have to be considered for all edges $e = (v_{i_1}, v_{i_2})$. In addition, the starting time T_i^S of a node is restricted to the $ASAP$[12]/ $ALAP$[13]-range (eq. 4.28) of v_i which is computed in a preprocessing step.

$$\forall v_i \in V^I : \quad T_i^D = \sum_{k \in KH} X_{i,k} * c_{i,k}^{th} + \sum_{k \in KP} Y_{i,k} * c_{i,k}^{ts} \tag{4.22}$$

$$\forall v_i \in V^{RI} \cup V^R : \quad T_i^D = \sum_{k \in KB} z_{i,k} * c_{i,k}^{tr} \tag{4.23}$$

$$\forall v_i \in V^{WO} \cup V^W : \quad T_i^D = \sum_{k \in KB} z_{i,k} * c_{i,k}^{tw} \tag{4.24}$$

$$\forall v_i \in V : \quad T_i^E = T_i^S + T_i^D \tag{4.25}$$

$$\forall v_i \in V : \quad T_i^E \leq C^t \leq MAX^t \tag{4.26}$$

$$\forall e = (v_{i_1}, v_{i_2}) \in E : \quad T_{i_2}^S \geq T_{i_1}^E \tag{4.27}$$

$$\forall v_i \in V : \quad ASAP(v_i) \leq T_i^S \leq ALAP(v_i) \tag{4.28}$$

[12] ASAP: As Soon As Possible
[13] ALAP: As Last As Possible

4.8.4 Hardware Sharing

If hardware sharing is considered, then it is not sufficient to model bindings between computation nodes $v_i \in V^I$ and hardware components h_k with the help of the binary variable $X_{i,k}$. In order to consider hardware sharing, the binding of v_i to the j-th hardware instance (of system component $c_l = I(v_i)$) contained in h_k has to be modelled using the binary binding variable $x_{i,j,k}$ (eq. 4.29). An example of hardware sharing is given in figure 4.30.

Example 11:

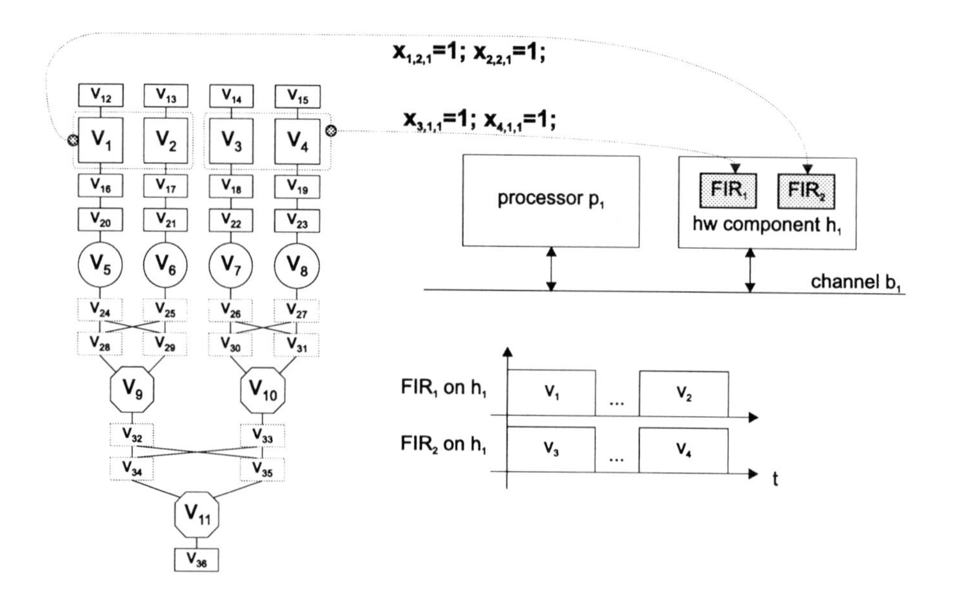

Figure 4.30. Hardware sharing

In figure 4.30, an example of sharing hardware resources is given to minimize the amount of chip area. The computation nodes v_3 and v_4, representing instances of an FIR-filter (system component c_1), are mapped to a first hardware instance $\mathtt{FIR}_1$ ($= h_{1,1,1}$) on hardware component h_1. Nodes v_1 and v_2 share the second hardware instance $\mathtt{FIR}_2$ ($= h_{2,1,1}$). Therefore, $x_{3,1,1} = x_{4,1,1} = 1$ and $x_{1,2,1} = x_{2,2,1} = 1$. Summarizing, four nodes are implemented by two hardware instances of FIR-filters on ASIC h_1. The timing diagram shows that v_1, v_2 and also v_3, v_4 have to be scheduled, but both hardware instances $\mathtt{FIR}_1$ and $\mathtt{FIR}_2$ are able to work concurrently.

A node v_i is implemented on h_k if it is bound to the j-th hardware instance of system component $c_l = I(v_i)$ on h_k (eq. 4.30) indicated by variable $x_{i,j,k}$. If at least one node v_i of system component c_l is bound to the j-th hardware instance of c_l on h_k, then $nx_{j,l,k} = 1$ (eq. 4.31, 4.32). The number $NX_{l,k}$ of used hardware instances of c_l on h_k is calculated by accumulating the variables $nx_{j,l,k}$ (eq. 4.33).

$$\forall k \in KH : \forall l \in L : N = \mid instances\_of(c_l) \mid :$$

$$\forall v_i \in V^I : I(v_i) = c_l : \forall j \in \{1, \ldots, N\} : \quad x_{i,j,k} \leq 1 \tag{4.29}$$

$$\forall v_i \in V^I : I(v_i) = c_l : \quad X_{i,k} = \sum_{j=1}^{N} x_{i,j,k} \tag{4.30}$$

$$\forall v_i \in V^I : I(v_i) = c_l : \forall j \in \{1, \ldots, N\} : \quad nx_{j,l,k} \geq x_{i,j,k} \tag{4.31}$$

$$\forall j \in \{1, \ldots, N\} : \quad nx_{j,l,k} \leq \sum_{v_i \in V^I : I(v_i) = c_l} x_{i,j,k} \tag{4.32}$$

$$NX_{l,k} = \sum_{j=1}^{N} nx_{j,l,k} \tag{4.33}$$

If hardware sharing is not considered, then equation 4.15 is used instead of equations 4.29-4.33.

4.8.5 *Interfacing*

Example 12:

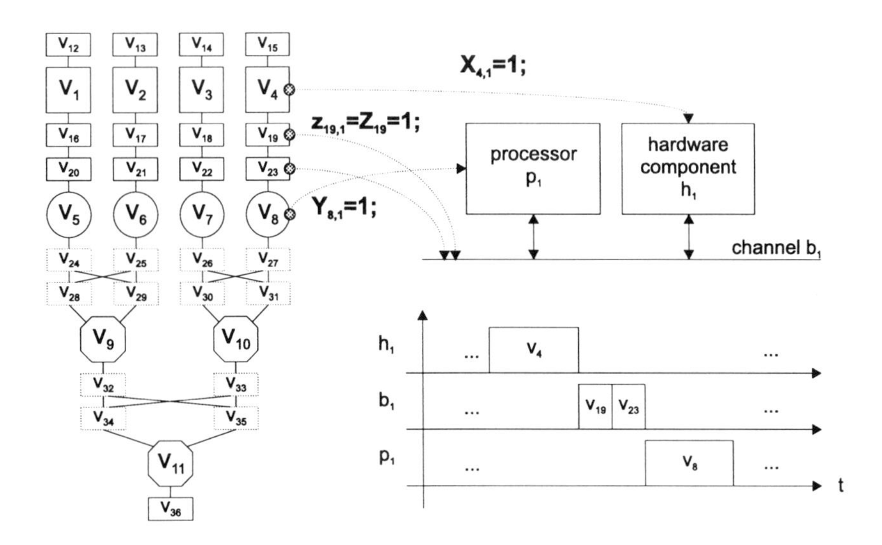

Figure 4.31. Interfacing

In figure 4.31, an example of a required interface is given . Node v_4 has been mapped to hardware component h_1 ($X_{4,1} = 1$) and v_8 to processor p_1 ($Y_{8,1} = 1$). Therefore, the output data of v_4, calculated on h_1 has to be transfered to v_8, implemented on p_1. This is done in two steps: First, the result computed by v_4 is written in a `WRITE`-phase (v_{19}) using communication channel b_1 ($z_{19,1} = 1$). Then, v_8 reads in a `READ`-phase (v_{23}) the data using communication channel b_1 ($z_{23,1} = 1$). The timing diagram shows that v_8 starts after v_4 has finished, data has been written by v_{19} and read by v_{23} using b_1.

An interface is required between two nodes $v_{i_1}, v_{i_2} \in V^I$ exchanging data ($v_{i_1} \in$ *uses_data_of*(v_{i_2})) if they are mapped to different processing units (eq. 4.35-4.39). In such a case, the **WRITE**- (v_{i_w}) and **READ**-nodes (v_{i_r}) between v_{i_1} and v_{i_2} have to indicate the need of an interface by $Z_{i_w} = Z_{i_r} = 1$. With the help of the interface binding variables $z_{i_w,k}$ and $z_{i_r,k}$, a channel b_k is selected implementing the data transfer from v_{i_1} to v_{i_2} (eq. 4.41). It is obvious that always $Z_{i_w} = Z_{i_r}$ and $z_{i_w,k} = z_{i_r,k}$. For this reason, all decision variables Z_{i_r} and $z_{i_r,k}$ for the **READ**-nodes ($v_{i_r} \in V^R$) are not necessary, because the variables Z_{i_w} and $z_{i_w,k}$ for the appropriate **WRITE**-node $v_{i_w} \in V^W$ represent the interface. Therefore, the variables Z_{i_r} and $z_{i_r,k}$ for the **READ**-nodes are not used in the MILP models generated by COOL. But for the sake of simplicity, they are used in this explanation.

$\forall v_{i_1}, v_{i_2} \in V^I : v_{i_1} \in uses\_data\_of(v_{i_2})$:
Let $(v_{i_1}, v_{i_w}, v_{i_r}, v_{i_2})$ be the path from v_{i_1} to v_{i_2} with $v_{i_w} \in V^W$ and $v_{i_r} \in V^R$.

$$Z_{i_w} \leq 1 \quad (4.34)$$

$$\forall k \in KH : \quad Z_{i_w} \geq X_{i_1,k} - X_{i_2,k} \quad (4.35)$$

$$\forall k \in KH : \quad Z_{i_w} \geq X_{i_2,k} - X_{i_1,k} \quad (4.36)$$

$$\forall k \in KP : \quad Z_{i_w} \geq Y_{i_1,k} - Y_{i_2,k} \quad (4.37)$$

$$\forall k \in KP : \quad Z_{i_w} \geq Y_{i_2,k} - Y_{i_1,k} \quad (4.38)$$

$$Z_{i_w} \rightarrow minimize \quad (4.39)$$

$$\forall k \in KB : \quad z_{i_w,k} \leq 1 \quad (4.40)$$

$$Z_{i_w} = \sum_{k \in KB} z_{i_w,k} \quad (4.41)$$

Equations 4.35-4.41 describe the interface constraints for communication nodes $v_i \in V^W \cup V^R$. Communication nodes for reading inputs and writing outputs always require additional communication, because inputs are written into memory and outputs are read from memory in COOL. This fact is modelled by equations 4.42-4.44.

$\forall v_i \in V^{RI} \cup V^{WO}$:

$$Z_i \leq 1 \quad (4.42)$$

$$\forall k \in KB : \quad z_{i,k} \leq 1 \quad (4.43)$$

$$Z_i = \sum_{k \in KB} z_{i,k} \quad (4.44)$$

If an interface is required, the additional communication times for `READ`-nodes and `WRITE`-nodes are computed by equations 4.45 and 4.46.

$$\forall v_i \in V^{RI} \cup V^R : \quad T_i^D = \sum_{k \in KB} z_{i,k} * c_{i,k}^{tr} \qquad (4.45)$$

$$\forall v_i \in V^{WO} \cup V^W : \quad T_i^D = \sum_{k \in KB} z_{i,k} * c_{i,k}^{tw} \qquad (4.46)$$

4.8.6 *Scheduling*

Scheduling has to be applied for nodes v_{i_1}, v_{i_2} if

- nodes $v_{i_1}, v_{i_2} \in V^I$ share the same hardware instance on the same ASIC, or
- nodes $v_{i_1}, v_{i_2} \in V^I$ are executed on the same processor, or
- nodes $v_{i_1}, v_{i_2} \in V^{RW}$ represent interfaces and share the same communication channel.

To sequentialize two nodes v_{i_1} and v_{i_2}, the binary decision variable b_{i_1,i_2} is used.

Example 13:

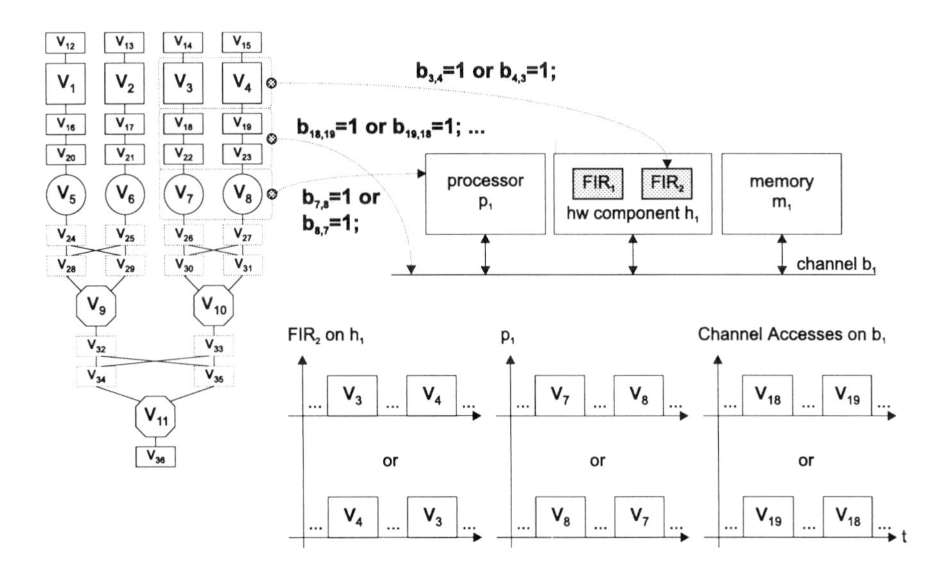

Figure 4.32. Application of scheduling constraints

In figure 4.32, all cases are depicted in which scheduling variables and constraints are required.

1. Nodes v_3 and v_4 are mapped to the same hardware instance of an FIR-filter on h_1: Therefore, v_3 and v_4 have to be scheduled.
2. Nodes v_7 and v_8 are mapped to the same processor p_1: Then, nodes v_7 has to be executed before v_8 ($b_{7,8} = 1$) or v_8 before v_7 ($b_{8,7} = 1$).
3. Nodes v_{18} and v_{19} are necessary to implement the required interface between v_4 (mapped to h_1) and v_8 (mapped to p_1).

To ease the description of the constraints required for scheduling, first a simplified model will be presented. This model proposes that nodes can be executed in sequence on the same resource without considering the channel accesses during READ- and WRITE-routines (see chapter 4.3.1.2).

Simplified Model without Consideration of Channel Accesses:. In this model, two nodes v_{i_1}, v_{i_2} are scheduled (if necessary) in such a way that either v_{i_2} starts after v_{i_1} finishes or so that v_{i_1} starts after v_{i_2} finishes (see figure 4.33).

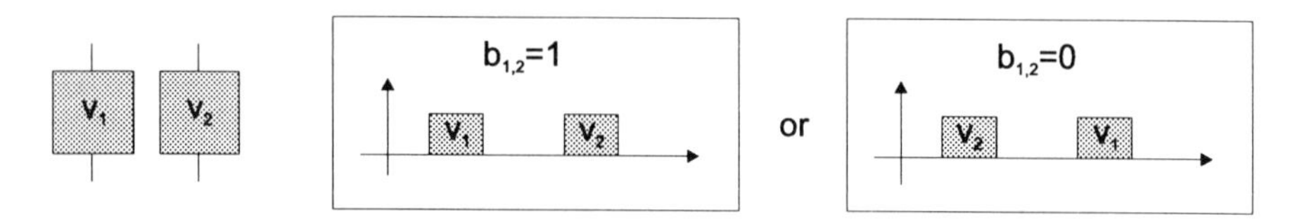

Figure 4.33. Simplified scheduling model

With the help of the binary decision variable b_{i_1,i_2} and the following constraints, it is possible to schedule two nodes v_{i_1}, v_{i_2} on a processor (eq. 4.47, 4.48), on a hardware instance (eq. 4.49, 4.50) or on a communication channel (eq. 4.51, 4.52):

$\forall k \in KP, \forall v_{i_1}, v_{i_2} \in V^I : v_{i_1} \in schedule\_nodes(v_{i_2})$

$$T_{i_1}^E \leq T_{i_2}^S + (3 - b_{i_1,i_2} - Y_{i_1,k} - Y_{i_2,k}) * C_1 \quad (4.47)$$

$$T_{i_2}^E \leq T_{i_1}^S + (2 + b_{i_1,i_2} - Y_{i_1,k} - Y_{i_2,k}) * C_2 \quad (4.48)$$

$\forall k \in KH :$
$\quad \forall l \in L : N = \mid instances\_of(c_l) \mid \geq 2 :$
$\quad\quad \forall j \in \{1, \ldots, N\} :$
$\quad\quad\quad \forall v_{i_1}, v_{i_2} \in V^I : v_{i_1}, v_{i_2} \in instances\_of(c_l) \wedge v_{i_1} \in schedule\_nodes(v_{i_2})$

$$T_{i_1}^E \leq T_{i_2}^S + (3 - b_{i_1,i_2} - x_{i_1,j,k} - x_{i_2,j,k}) * C_3 \quad (4.49)$$

$$T_{i_2}^E \leq T_{i_1}^S + (2 + b_{i_1,i_2} - x_{i_1,j,k} - x_{i_2,j,k}) * C_4 \quad (4.50)$$

$\forall k \in KB, \forall v_{i_1}, v_{i_2} \in V^{RW} : v_{i_1} \in schedule\_nodes(v_{i_2})$

$$T^E_{i_1} \leq T^S_{i_2} + (3 - b_{i_1,i_2} - z_{i_1,k} - z_{i_2,k}) * C_5 \quad (4.51)$$
$$T^E_{i_2} \leq T^S_{i_1} + (2 + b_{i_1,i_2} - z_{i_1,k} - z_{i_2,k}) * C_6 \quad (4.52)$$

The basic idea of these constraints is always the same. For this reason, only the constraints 4.47 and 4.48 for scheduling nodes v_{i_1} and v_{i_2} using the same processors p_k are described in the following. If v_{i_1} and v_{i_2} have to be scheduled ($Y_{i_1,k} = Y_{i_2,k} = 1$), then one of the following conditions has to be fulfilled:

1. v_{i_1} is executed before v_{i_2} ($b_{i_1,i_2} = 1$) $\Rightarrow T^E_{i_1} \leq T^S_{i_2}$, or
2. v_{i_2} is executed before v_{i_1} ($b_{i_1,i_2} = 0$) $\Rightarrow T^E_{i_2} \leq T^S_{i_1}$.

This fact is modelled by the constraints defined in equations 4.47 and 4.48:

$Y_{i_1,k} = Y_{i_2,k} = 1$	b_{i_1,i_2}	equation 4.47	equation 4.48
yes	0	$T^E_{i_1} \leq T^S_{i_2} + C_1$	$T^E_{i_2} \leq T^S_{i_1}$
yes	1	$T^E_{i_1} \leq T^S_{i_2}$	$T^E_{i_2} \leq T^S_{i_1} + C_2$
no	0, 1	$T^E_{i_1} \leq T^S_{i_2} + n_1 * C_1$, $n_1 \geq 1$	$T^E_{i_2} \leq T^S_{i_1} + n_2 * C_2$, $n_2 \geq 1$

If $Y_{i_1,k} = Y_{i_2,k} = 1$, only one of equations 4.47 and 4.48 results in a hard constraint. If $b_{i_1,i_2} = 0$, equation 4.47 has no effect and if $b_{i_1,i_2} = 1$ equation 4.48 can be ignored. If either $Y_{i_1,k} = 0$ or $Y_{i_2,k} = 0$, both constraints have no effect if C_1 and C_2 are dimensioned correctly. It can be shown that C_1 and C_2 have the following lower bounds:

Let $MaxExecTime(v_i) = Max\left\{\{c^{ts}_{i,k_1} \mid k_1 \in KP\} \cup \{c^{th}_{i,k_2} \mid k_2 \in KH\}\right\}$

1. $C_1 = \lceil ALAP(v_{i_1}) + MaxExecTime(v_{i_1}) - ASAP(v_{i_2}) \rceil$, because

$$\begin{aligned} T^E_{i_1} &\leq T^S_{i_2} + C_1 \\ &\leq T^S_{i_2} + ALAP(v_{i_1}) + MaxExecTime(v_{i_1}) - ASAP(v_{i_2}) \\ &\leq ALAP(v_{i_1}) + MaxExecTime(v_{i_1}) \qquad \square \end{aligned}$$

2. $C_2 = \lceil ALAP(v_{i_2}) + MaxExecTime(v_{i_2}) - ASAP(v_{i_1}) \rceil \ldots \square$

Precise Model with Consideration of Channel Accesses:. The simplified approach described before suffers from not considering channel accesses during `READ`- and `WRITE`-routines. In chapter 4.1.4, it has been shown that no computation can be executed on t_k during a channel access of processing unit $t_k \in \mathcal{PU}$. A node v_{i_2} may be started after v_{i_1} on the same processor (ASIC) if v_{i_1} has written all its results required by other processing units.

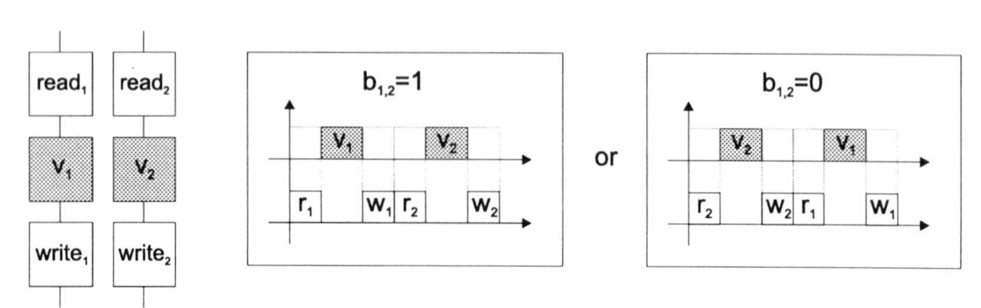

Figure 4.34. Precise scheduling model

For this reason, if a node v_{i_2} is scheduled after a node v_{i_1} on a processor (eq. 4.53), then v_{i_2} may start after all data produced by v_{i_1} has been written (see figure 4.34). The precise model defines the timing behavior of computation nodes with equations 4.53-4.56:

$\forall k \in KP, \forall v_{i_1}, v_{i_2} \in V^I : v_{i_1} \in schedule\_nodes(v_{i_2})$

$\forall v_{i'_1} \in write\_nodes(v_{i_1}), \forall v_{i'_2} \in read\_nodes(v_{i_2}) :$

$$T^E_{i'_1} \leq T^S_{i'_2} + (3 - b_{i_1,i_2} - Y_{i_1,k} - Y_{i_2,k}) * C'_1 \tag{4.53}$$

$\forall v_{i'_2} \in write\_nodes(v_{i_2}), \forall v_{i'_1} \in read\_nodes(v_{i_1}) :$

$$T^E_{i'_2} \leq T^S_{i'_1} + (2 + b_{i_1,i_2} - Y_{i_1,k} - Y_{i_2,k}) * C'_2 \tag{4.54}$$

$\forall k \in KH :$
$\quad \forall l \in L : N = | instances\_of(c_l) | \geq 2 :$
$\quad\quad \forall j \in \{1, \ldots, N\} :$
$\quad\quad\quad \forall v_{i_1}, v_{i_2} \in V^I : v_{i_1}, v_{i_2} \in instances\_of(c_l) \wedge v_{i_1} \in schedule\_nodes(v_{i_2})$

$\forall v_{i'_1} \in write\_nodes(v_{i_1}), \forall v_{i'_2} \in read\_nodes(v_{i_2}) :$

$$T^E_{i'_1} \leq T^S_{i'_2} + (3 - b_{i_1,i_2} - x_{i_1,j,k} - x_{i_2,j,k}) * C'_3 \tag{4.55}$$

$\forall v_{i'_2} \in write\_nodes(v_{i_2}), \forall v_{i'_1} \in read\_nodes(v_{i_1}) :$

$$T^E_{i'_2} \leq T^S_{i'_1} + (2 + b_{i_1,i_2} - x_{i_1,j,k} - x_{i_2,j,k}) * C'_4 \tag{4.56}$$

Equations 4.53-4.56 of the precise model replace equations 4.47-4.50 of the simplified model. The constants $C'_1, \ldots, C'_4$ can be computed corresponding to the computation of $C_1, \ldots, C_4$ in the simplified model.

4.9 HW/SW Partitioning based on Heuristic Scheduling

Resource constrained scheduling is an NP-complete problem [GaJo79]. Therefore, solving the hardware/software partitioning problem optimally cannot be efficiently done, because resource constraint scheduling is one key aspect of it. In figure 4.35, the number of variables for several MILP models are depicted. This figure has been generated from a set of different examples partitioned with COOL.

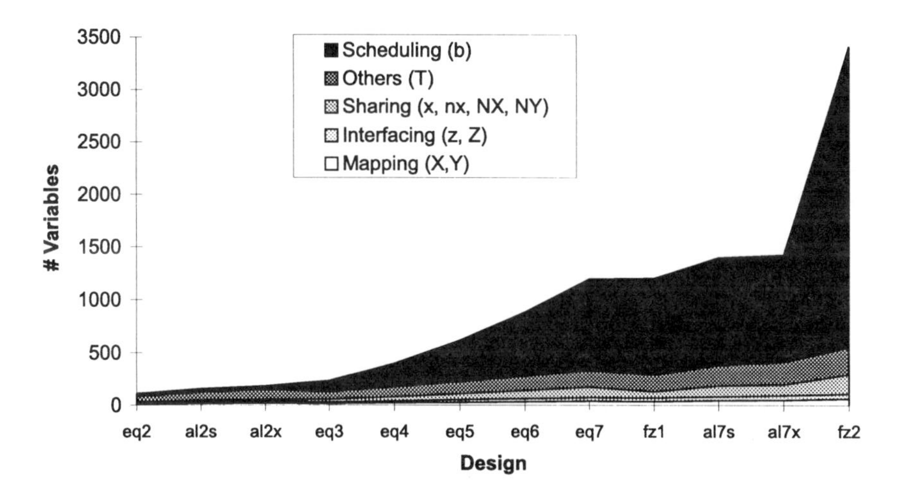

Figure 4.35. Number of variables for generated MILP models

It can be seen that the b-variables (required to model scheduling constraints) represent the major part of the total number of variables. Scheduling complicates the hardware/software partitioning problem drastically.

For this reason, in addition to the optimal approach (`P_MILP`) presented in section 4.8, another partitioning algorithm has been developed, dividing the hardware/software partitioning problem into two subproblems: a *mapping problem* and a *scheduling problem*. This algorithm is called *partitioning based on heuristic scheduling* (`P_HS`). `P_HS` solves both aspects of the problem sequentially using a combination of mixed integer linear programming and a heuristic while iterating the following steps:

Step I: Solve an MILP model for the hardware/software mapping with the help of approximated time values. This step determines on which processing unit a node will be executed. The execution order on each processing unit itself is not computed.

Step II: Compute a schedule for the hardware/software mapping of step I, either by

- using a second MILP model or
- using a constructive algorithm (e.g. *list scheduling*).

Step III: If the resulting total execution time of step II violates the timing constraint, repeat the first two steps with a refined timing constraint that is tighter than the approximated total time of step III (see figure 4.36).

The following example illustrates the idea of the algorithm.

Example 14:

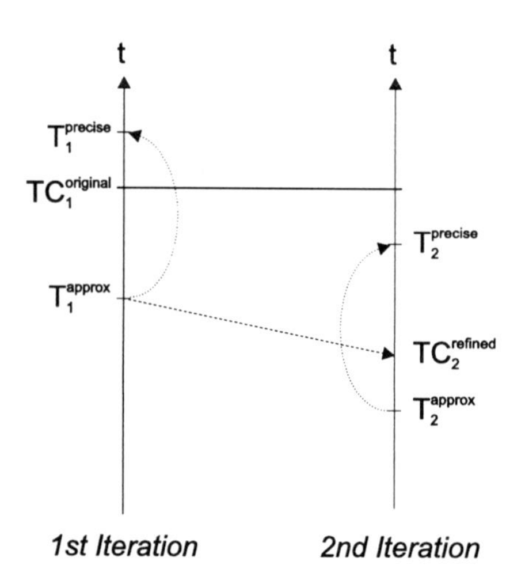

Figure 4.36. Heuristic scheduling

1st iteration: The first step computes a mapping and an approximated system execution time T_1^{approx} which fulfills the given timing constraint $TC_1^{original}$. However, the precise system execution time $T_1^{precise}$ violates $TC_1^{original}$. For this reason, a second iteration with a refined timing constraint $TC_2^{refined}$ (which has to be smaller than T_1^{approx} to prevent that the last mapping is computed again) is necessary.

2nd iteration: Step 1 computes again a mapping and a new approximated system execution time $T_2^{approx} \leq TC_2^{refined}$. In this case, a precise execution time $T_2^{precise}$ is computed in step 2 fulfilling the original timing constraint $TC_1^{original}$. Therefore, the second mapping represents the solution.

In the following, step 1 and 2 will be described in more detail and different approaches to compute a schedule for a given hardware/software mapping will be presented.

4.9.1 Step I: Mapping with Approximated Schedule

The MILP model generated in this step contains nearly all constraints presented in the section before. But all b_{i_1,i_2}-variables and constraints modelling scheduling (eq. 4.47-4.56) are missing. The goal is to compute a hardware/software mapping without determining a precise schedule by approximating the schedule times of all nodes. In addition to the data-dependency and ASAP/ALAP constraints (eq. 4.22-4.28), additional constraints based on predecessor and dominator nodes are used to approximate time values more precisely:

Predecessor nodes: A node v is ready to start if all its *predecessors* have finished their execution. The effect of being forced to schedule some of these predecessor nodes can be exploited to estimate the starting time of v.

Example 15:

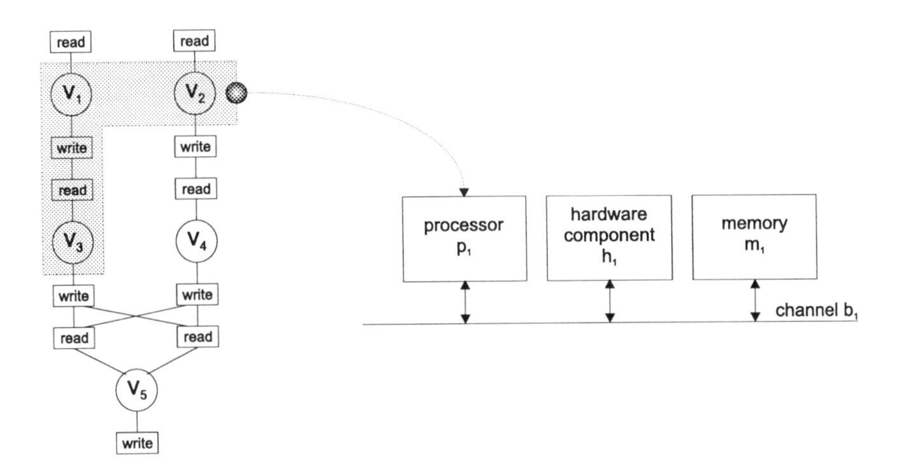

Figure 4.37. Using predecessor nodes in heuristic scheduling

In figure 4.37, the starting time of v_5 is at least the sum of execution times of v_1, v_2, v_3, because all three nodes have been mapped to p_1 and have to be scheduled.

$\Rightarrow$ $$T_5^S \geq c_{1,1}^{ts} + c_{2,1}^{ts} + c_{3,1}^{ts}.$$

The starting time of a node v_i is equal to or greater than the accumulated software execution times of all predecessor nodes v_i (eq. 4.57) on a processor p_k. Similar constraints can be added if hardware sharing (eq. 4.58) and/or interfacing (eq. 4.59) are considered.

$\forall v_{i_1} \in V$:
Let $LV = pred\_nodes(v_{i_1}) \cap V^I$
$LW = pred\_nodes(v_{i_1}) \cap (V^{WO} \cup V^W)$
$LR = pred\_nodes(v_{i_1}) \cap (V^{RI} \cup V^R)$

$\forall v_{i_1} \in V^I : \forall k \in KP$:

$$T_{i_1}^S \geq \sum_{v_{i_2} \in LV} Y_{i_2,k} * c_{i_2,k}^{ts} \tag{4.57}$$

$\forall l \in L : N = | instances\_of(c_l) | \geq 2$:
$\forall j \leq N : \forall v_{i_1} \in V^I : \forall k \in KH$:

$$T_{i_1}^S \geq \sum_{v_{i_2} \in LV, I(v_{i_2}) = c_l} x_{i_2,j,k} * c_{l,k}^{th} \tag{4.58}$$

$\forall v_{i_1} \in V^{RW} : \forall k \in KB$:

$$T_{i_1}^S \geq \sum_{v_{i_2} \in LW} z_{i_2,k} * c_{i_2,k}^{tw} + \sum_{v_{i_2} \in LR} z_{i_2,k} * c_{i_2,k}^{tr} \tag{4.59}$$

Dominator nodes: Another possibility to estimate the starting time of a node v is to consider dominator nodes. A *dominator node* w of v is a node, for which each path to v contains w (see definition 11). v can be started if dominator w and all nodes between w and v have been executed.

Example 16:

In figure 4.38, the starting time of v_6 is at least the sum of the ending time of v_3 and the execution times of v_4, v_5, because v_4, v_5 have to be scheduled after executing v_3.
$\Rightarrow \quad T_6^S \geq T_3^E + c_{4,1}^{ts} + c_{5,1}^{ts}$.

The starting time of a node v_{i_1} is equal to or greater than the sum of the ending time of the dominator node v_{i_0} of v_{i_1} and the software execution times on processor p_k of all nodes on the paths between v_{i_0} and v_{i_1} (eq. 4.60). Equation 4.61 defines the same constraint for the hardware execution times of all shared nodes on the paths between v_{i_0} and v_{i_1}. Equation 4.62 defines the equivalent constraint considering communication times of required interfaces.

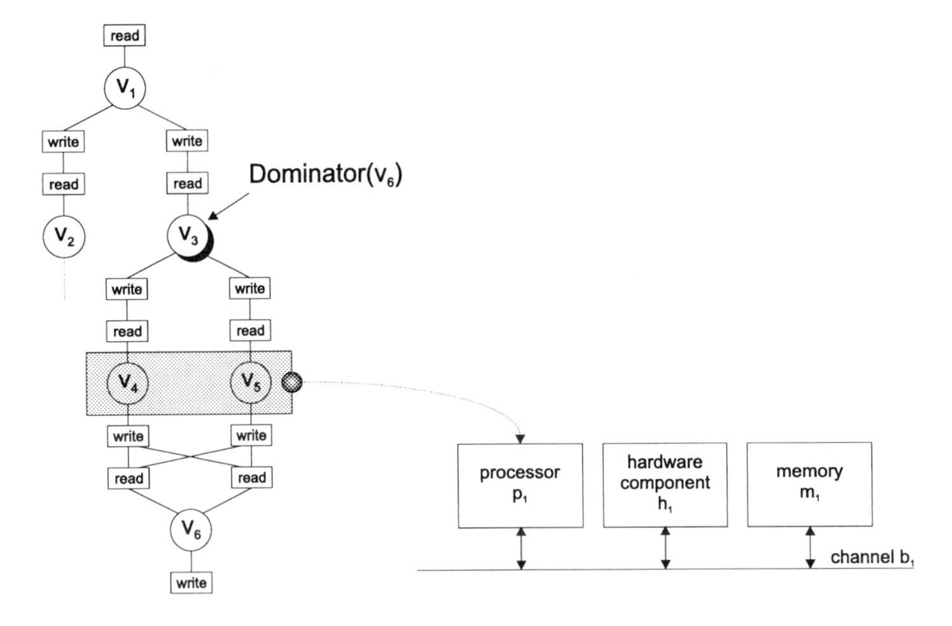

Figure 4.38. Using dominator nodes in heuristic scheduling

$\forall v_{i_0}, v_{i_1} \in V : v_{i_0} \in dominator\_nodes(v_{i_1}) :$

Let $LV = pred\_nodes(v_{i_1}) \cap succ\_nodes(v_{i_0}) \cap V^I$

$LW = pred\_nodes(v_{i_1}) \cap succ\_nodes(v_{i_0}) \cap (V^{WO} \cup V^W)$

$LR = pred\_nodes(v_{i_1}) \cap succ\_nodes(v_{i_0}) \cap (V^{RI} \cup V^R)$

$\forall v_{i_1} \in V^I : \forall k \in KP :$

$$T^S_{i_1} \geq T^E_{i_0} + \sum_{v_{i_2} \in LV} Y_{i_2,k} * c^{ts}_{i_2,k} \tag{4.60}$$

$\forall l \in L : N = \mid instances\_of(c_l) \mid \geq 2 :$

$\forall j \leq N : \forall v_{i_1} \in V^I : \forall k \in KH :$

$$T^S_{i_1} \geq T^E_{i_0} + \sum_{v_{i_2} \in LV, I(v_{i_2}) = c_l} x_{i_2,j,k} * c^{th}_{l,k} \tag{4.61}$$

$\forall v_{i_1} \in V^{RW} : \forall k \in KB :$

$$T^S_{i_1} \geq T^E_{i_0} + \sum_{v_{i_2} \in LW} z_{i_2,k} * c^{tw}_{i_2,k} + \sum_{v_{i_2} \in LR} z_{i_2,k} * c^{tr}_{i_2,k} \tag{4.62}$$

After a hardware/software mapping has been calculated while estimating the schedule, the goal is to compute a precise schedule for this mapping in the second step.

4.9.2 *Step II: Scheduling solved by MILP*

The first possibility of solving the scheduling problem is to generate a second MILP problem, consisting only of the constraints required for scheduling. The overall optimization goal of the hardware/software partitioning approach is hardware minimization under timing constraints. In this case, it can be sufficient to find a solution for the scheduling problem encoded in an MILP model. Therefore, there are two possibilities to compute a schedule using MILP:

1. `S_MILP_OPT` solves the MILP model as an *optimization problem* (with an objective function including system execution time).
2. `S_MILP_CS` solves the MILP model as a *constraint-satisfaction problem* (with a constant objective function).

In both cases, the variables and the constraints are the same but the objective functions are different. But nevertheless, solving the scheduling problem using MILP has always the great disadvantage that the computation time may exponentially grow. Therefore, a second scheduling approach has been implemented.

4.9.3 *Step II: Scheduling solved by List Scheduling*

The scheduling problem can also be solved by constructive heuristic algorithms. Algorithm 2 represents a special implementation (called `S_LS` in the following) of the well-known *list scheduling* algorithm. For a given partitioning graph $G^P = (V, E, C, I)$ with $n_V =| V |$ and $n_E =| E |$, `S_LS` computes such a schedule in time $O(n_V^2)$.

Algorithm 2: ExtendedListScheduling (S_LS)

```
(1)  algorithm ExtendedListScheduling;
(2)      input G^P : PartitioningGraph; heuristic: natural;
(3)      output C^t : real;
(4)  {
(5)      variable G^S = (V',E',C',I') : PartitioningGraph;
(6)      variable i,k,n               : natural;
(7)      variable r_k                 : Resource;
(8)      variable v_i,w               : node;
(9)      variable L_not_scheduled     : list of node;
(10)     variable asapalap            : node_array of real;
(11)     variable T_i^S,T_i^D,T_i^E,t : real;
(12)
(13)     G^S = Remove_Unnecessary_Communication_Nodes(G^P);
(14)     asapalap = Compute_ASAP_ALAP(G^S);
(15)     L_not_scheduled = Topsort_With_Heuristic(G^S, asapalap, heuristic);
(16)     C^t = 0;
(17)     for i = 1 to | V' | do
(18)         T_i^S = T_i^E = 0;
(19)     for n = 1 to | V' | do
(20)     {
(21)         v_i = GetNextNode(G^S, L_not_scheduled, asapalap);
(22)         r_k = Resource_of(v_i);
(23)         t = GetStartTime(GetPredecessors(G^S,v_i), ScheduleList(r_k));
(24)         T_i^S = t;
(25)         T_i^E = T_i^S + T_i^D;
(26)         C^t = Max(C^t,T_i^E);
(27)         ScheduleList(r_k).insert(v_i);
(28)         L_not_scheduled.delete(v_i);
(29)         if v_i ∈ V^R ∪ V^RI then
(30)             Add_new_Edges(G^S, v_i, L_not_scheduled);
(31)     }
(32) }
```

In a first step (line 13), all unnecessary communication nodes are eliminated from the partitioning graph G^P resulting in a new graph G^S. Then, $ASAP$[14] and $ALAP$[15] values are computed for G^S (line 14). This can be done using a DFS[16] graph algorithm in time $O(n_V + n_E)$. Afterwards *topological sorting* (line 15) sorts the nodes by using a user-defined heuristic, e.g. "sorting by increasing ALAP-values". The time complexity of `Topsort_With_Heuristic` is $O(n_V * log(n_V))$. After all timing variables have been initialized (lines 16 to 18), the timing values are updated by scheduling the nodes of G^S iteratively. In each iteration, a node v_i is determined which has to be scheduled next (line 21) by searching the first node v_i in *L_not_scheduled* which can be scheduled. For this node, the starting time is computed (lines 23 to 25) by considering the ending times of all predecessor nodes of v_i and the nodes which have been

[14] ASAP: As Soon As Possible
[15] ALAP: As Last As Possible
[16] DFS: Depth First Search

scheduled on the same resource r_k. Existing time slots in the local schedule of r_k are considered. Then, node v_i is inserted into the local schedule list of resource r_k (line 27) and deleted from *L_not_scheduled* (line 28). If v_i represents a **READ**-node, additional edges are inserted into the graph (lines 29 and 30) to guarantee the execution order presented in section 4.1.4.

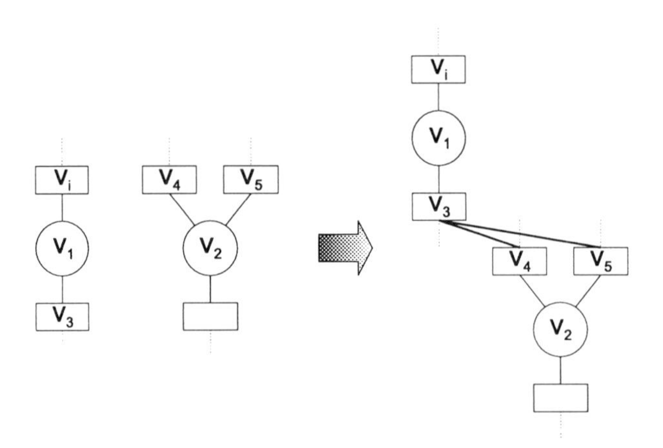

Figure 4.39. Additional edges

Figure 4.39 illustrates this step. Let $v_1, v_2 \in V^I$ be two computation nodes sharing the same resource. If **READ**-node v_i of v_1 is scheduled, two new edges $(v_3, v_4), (v_3, v_5)$ are created connecting the **WRITE**-nodes of v_1 to the **READ**-nodes of v_2. These edges prevent v_2 from being scheduled on r_k before v_1 has finished and written its results. After all iterations of **S_LS** have been executed, the schedule has been computed and the results stored in G^S are copied to the partitioning graph G^P. In figure 4.40, an example of the presented scheduling algorithm is given.

Example 17:

At the top of figure 4.40 the result of the hardware/software mapping phase is given for the 4-band equalizer. Nodes v_1 and v_2 have been mapped to a first instance of an FIR-filter in hardware, v_3 and v_4 to a second instance. All other computation nodes $(v_5, \ldots, v_{11})$ have been mapped to a processor. Therefore, the communication nodes $v_{24}, \ldots, v_{35}$ are not necessary and will be removed in step 13. The result of the topological sorting considering ALAP-values is also depicted in figure 4.40. Only $v_{12}, \ldots, v_{15}$ can be scheduled first illustrated by the gray color. Node v_{12} is selected in the first iteration, because it is the first node in list *L_not_scheduled* which can be scheduled. v_{12} represents a **READ**-node of communication node v_1 and v_1, v_2 share the same hardware resource. For this reason, an additional edge between the **WRITE**-node v_{16} of v_1 and the **READ**-node v_{13} of v_2 is inserted. This edge guarantees that v_{13} is scheduled after v_{16}. During the second iteration, v_{14} is selected and edge (v_{18}, v_{15}) is inserted. During the third iteration, the first computation node v_1 is scheduled. The result after executing all iterations is depicted at the bottom of figure 4.40.

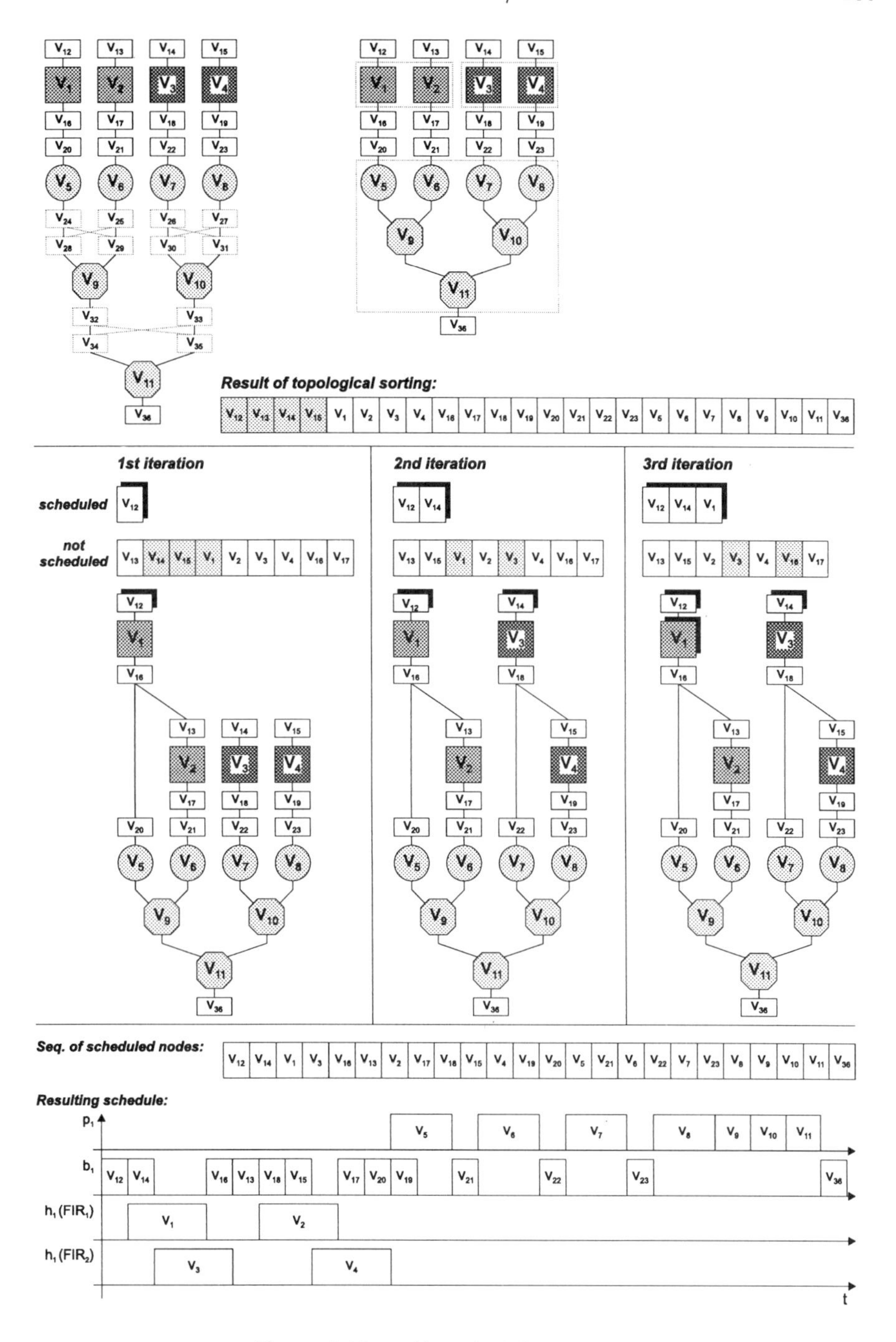

Figure 4.40. List scheduling approach

4.9.4 Results I: Scheduling Algorithms

In this section some results will be presented comparing different scheduling algorithms which have been described in section 4.9.3. For this purpose, a set of systems (see table 4.4) have been specified with COOL.

		System		Partitioning Graph	
Ex.	Description	#comp.	#inst.	#nodes	#edges
eq2	2-band equalizer	3	5	18	24
eq3	3-band equalizer	3	8	29	40
eq4	4-band equalizer	3	11	40	56
eq5	5-band equalizer	3	14	51	72
eq6	6-band equalizer	3	17	62	88
eq7	7-band equalizer	3	20	73	104
al1	a first audio system	6	8	30	42
al2	a second audio system	8	10	38	54
al3	a third audio system	6	23	90	137
al4	a fourth audio system	8	25	98	149
fz1	a first fuzzy controller	7	19	71	111
fz2	a second fuzzy controller	7	31	119	189

Table 4.4. Set of benchmarks

These systems are algorithms of the audio [PG245] and fuzzy logic [PG293b] domain. The equalizers `eq2...eq7` consist of digital filters, multipliers and adders. The first audio system `al1` contains a mixer (for four incoming audio samples), a fader for fading in or out the mixed signal, a 2-band equalizer and a balance ruler for the left and the right channel. In addition to `al1`, the second audio system `al2` includes an echo and a component simulating virtual boxes. In the audio systems `al3` and `al4`, the 2-band equalizer of `al1` and `al2` is replaced by a 7-band equalizer. The fuzzy controllers `fz1` and `fz2` have been developed to minimize the average latency of cars waiting at a crossing with three different tracks. They are described in more detail in [PG293b].

Most of the systems are small examples, but they are well suited to examine the algorithms for the following reasons:

- they have a high degree of concurrency and
- allow intensive hardware sharing effects, because most of the system components are multiply instantiated.

After specifying these systems, they have been partitioned with COOL onto a single-processor-single-ASIC target architecture. This has been done for a set of different timing constraints, computing an AT-curve as depicted in figure 4.41. All partitions have been computed on a Sun-SPARC-20 workstation.

The AT-curve is calculated by starting with a pure software solution (which is very slow in most cases). Then, the timing constraints are iteratively decreased

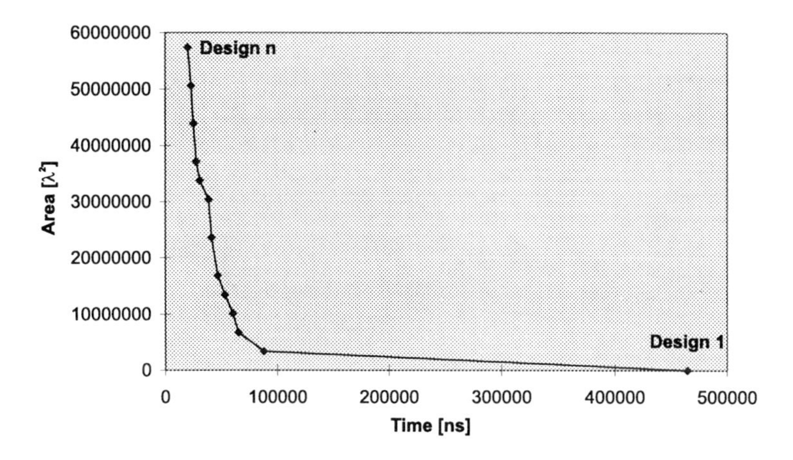

Figure 4.41. Computation of an AT-curve

resulting in more chip area until the fastest hardware/software partition has been computed (mostly a pure hardware solution).

For each computed hardware/software partition the following scheduling algorithms have been applied to compute a good (optimal) schedule:

- `S_LS`: list scheduling,
- `S_MILP_CS`: problem formulated as MILP problem and solved as a constraint satisfaction problem.
- `S_LS_MILP_CS`: MILP problem solved as a constraint satisfaction problem, but using the result of `S_LS` to approximate the ASAP/ALAP-values more precisely.
- `S_MILP_OPT`: problem formulated as MILP which is solved optimally.

In figure 4.42, the computed system execution times of all scheduling algorithms are depicted. To compare the different scheduling algorithms, the quality of the computed schedules and the required computation time will be compared.

Quality. The worst case deviation (see figure 4.43) between the optimal solution and the list scheduling solution has been 14.4%, the average deviation has been 0.93%. These results of `S_LS` are in both cases better than `S_MILP_CS`. The results of `S_LS` can be improved by performing `S_LS_MILP_CS`. The average (worst case) deviation can be decreased from 0.93% to 0.55% (from 14.4% to 10.1%). The main difference between `S_LS`, `S_MILP_CS` and `S_MILP_OPT` is the required computation time as depicted in figure 4.44

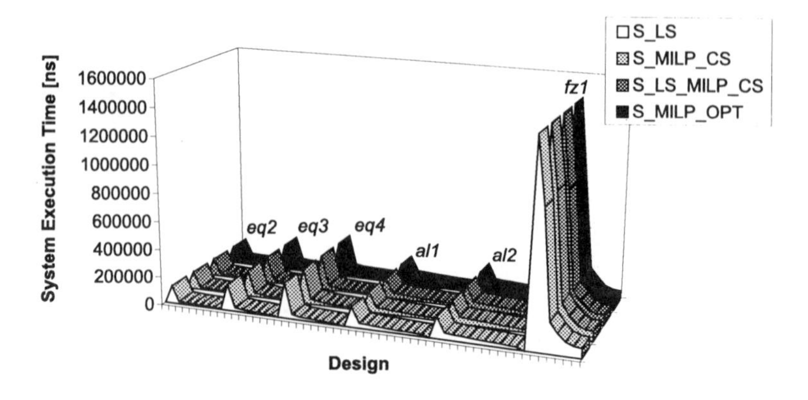

Figure 4.42. Scheduling results

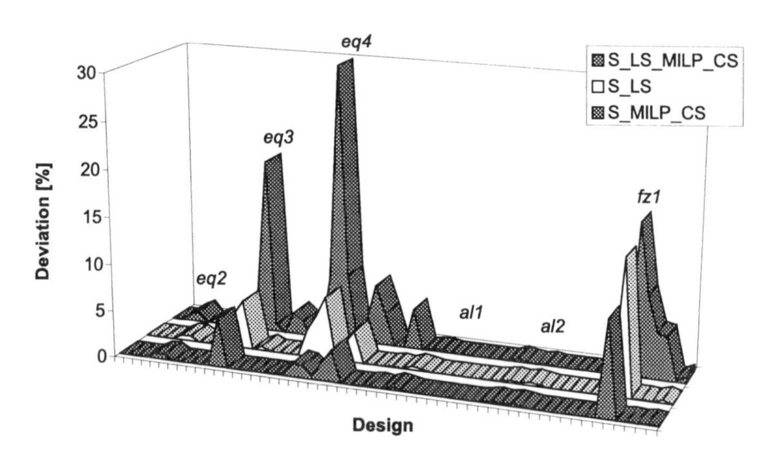

Figure 4.43. Scheduling precision

Computation Time. Whereas S_LS computes the results on the average in 0.5 seconds, S_MILP_CS needs 5.7 and S_MILP_OPT 41.7 seconds. The worst-case computation times are 1.7 seconds for S_LS, 119.9 seconds for S_MILP_CS and 1273.6 seconds for S_MILP_OPT. The average and worst case computation times of S_LS_MILP_CS and S_MILP_CS are nearly equal.

Table 4.5 summarizes the results obtained by comparing the scheduling algorithms for 50 hardware/software partitions of 6 different systems:

This difference in computation time grows dramatically if the systems become more complex. In such cases, solving the scheduling problem by MILP (either by S_MILP_CS or S_MILP_OPT) becomes impossible. Summarizing, it can be stated that S_LS is very well suited to compute a schedule for a given hard-

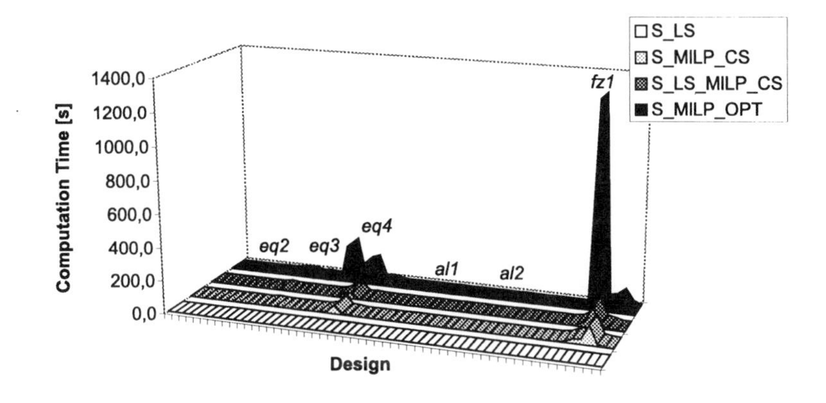

Figure 4.44. Computation time of scheduling algorithms

Example	Deviation [%] from optimum		Computation Time [sec]	
	average	worst-case	average	worst-case
S_LS	0.93	14.4	0.5	1.7
S_MILP_CS	2.53	29.8	5.7	119.9
S_LS_MILP_CS	0.55	10.1	5.5	118.8
S_MILP_OPT	0	0	41.7	1273.6

Table 4.5. Results of scheduling algorithms

ware/software mapping, because of its ratio between quality and computation time.

4.9.5 Results II: HW/SW Partitioning using Heuristic Scheduling

After the scheduling algorithms have been compared, now the results of the hardware/software partitioning algorithm based on *heuristic scheduling* will be presented using `S_LS` to compute a schedule for a hardware/software partition. Optimal solutions for the hardware/software partitioning problems can be computed by solving the MILP model including the scheduling constraints. This partitioning approach will be called `P_MILP` in the following. It has been shown during benchmarking that `P_MILP` can only be performed for smaller examples. The partitioning approach using heuristic scheduling, `P_HS`, has been evaluated by comparing the results of 35 designs for 4 different systems. The results are depicted in figure 4.45.

For 20 of these 35 partitions, `P_HS` has computed the optimal solution. The average deviation to the optimum has been about 1%, the worst case deviation about 18.4%. The computation time of `P_HS` (see table 4.6) has been drastically

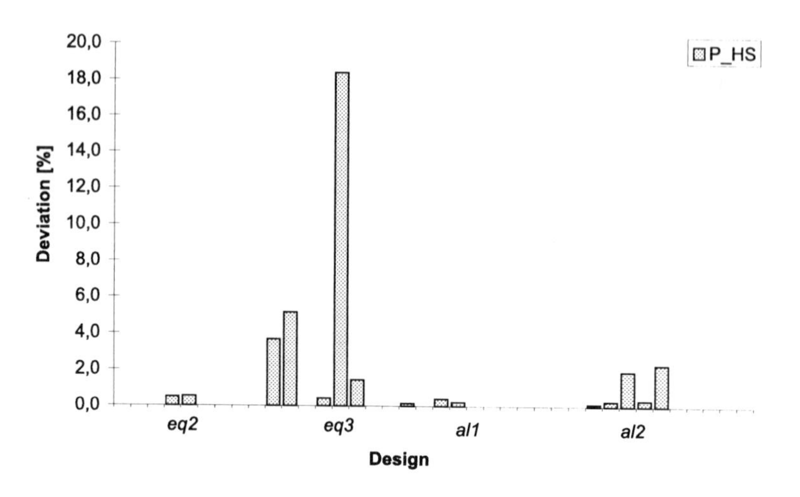

Figure 4.45. Deviation of computed hardware area

reduced compared to P_MILP, because the difficult scheduling aspect is now solved by list scheduling in short computation time. Therefore, the worst case (average) computation time has been reduced from 8837.3 (262.4) seconds for P_MILP to 16.4 (4.1) seconds for P_HS.

Example	Deviation [%] from optimum average	worst-case	Computation Time [sec] average	worst-case
P_HS	1.0	18.4	4.1	16.4
P_MILP	0	0	262.4	8837.3

Table 4.6. Results of partitioning algorithms

Summarizing, the optimal approach P_MILP can only be applied to systems with very small complexity or if the model is strongly simplified, for example,

- if some aspects are not considered (e.g. interfacing could be neglected for systems which are not communication-intensive) or
- if the designer defines additional constraints (e.g. mapping, scheduling or resource constraints).

In both cases the search space is restricted so that this reduced problem can hopefully be solved optimally in acceptable computation time.

In contrast, the heuristic scheduling approach can solve partitioning problems with a larger number of nodes. In figure 4.46, the results of P_HS for 104 designs of 12 different systems are depicted.

Clearly, the worst case computation time of P_HS also explodes in some cases. For this reason, another hardware/software partitioning approach has been

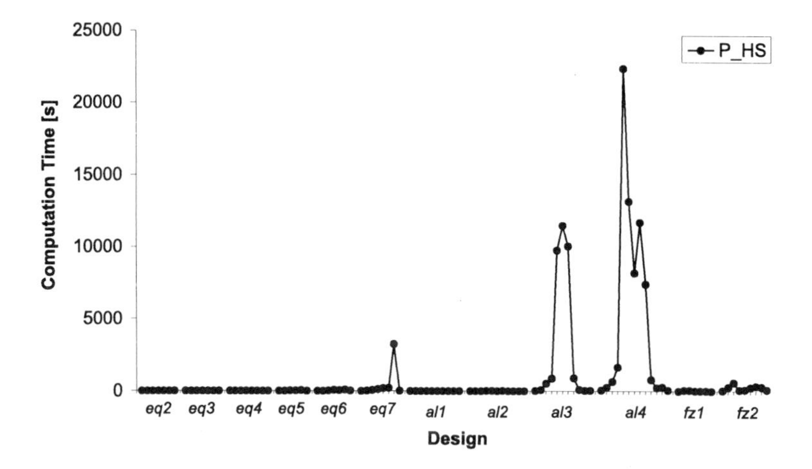

Figure 4.46. Computation time of heuristic scheduling partitioning algorithm

developed. It is based on genetic algorithms and will be presented in the following section.

4.10 HW/SW Partitioning based on Genetic Algorithms

In COOL, a second optimization approach is integrated based on *genetic algorithms.* The approach to solve the partitioning problem with GA will be called P_GA. It has been implemented using the C-library PGAPACK [Levi96] simplifying the implementation of genetic algorithms, because all genetic operators are fully implemented. In addition, genes are not only able to encode binary but also other discrete values. The flow-of-control is depicted in figure 4.47.

First, the user defines *genetic parameters* to steer the optimization process. Then, the value ranges for the genes are defined and the initial population P_0 is computed with a population size p. P_0 is computed randomly, but optionally it is also possible to define some start solutions to speed up the optimization process, e.g.

- a cheap solution (all nodes are implemented with minimal costs),
- a fast solution (all nodes are implemented using the fastest implementation alternative),
- low-communication solutions (all nodes are implement on the same processing unit)

It came out during benchmarking that finding solutions for hard timing constraints is difficult. Therefore, in particular, starting with a fast solution is very helpful.

After this initialization phase, the genetic algorithm iteratively improves the populations. For each generation i, all chromosomes of the population are directly evaluated by the *fitness function*. It will be shown that no *repair mechanism* is necessary because of the encoding presented in section 4.10.2. The genetic algorithm developed for hardware/software partitioning is similar to the partitioning approach based on heuristic scheduling which has been described in section 4.9. Mapping and binding are separated from scheduling. The chromosomes encode the mapping and binding, the scheduling is computed heuristically using the list scheduling algorithm (S_LS) presented in section 4.9.3. Then, the selection and variation phases, including crossover and mutation, compute a new population for each generation steered by the initial user-defined parameters. After the populations of all generations have been computed or a user-defined stopping criterion is fulfilled, the genetic algorithm stops and returns the chromosome representing the best solution.

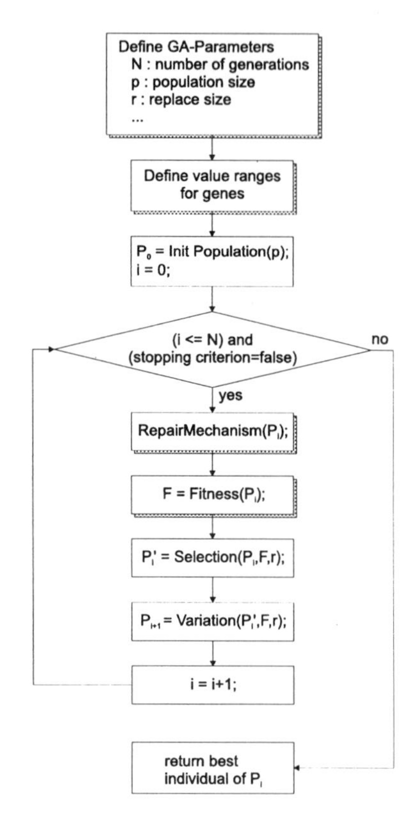

Figure 4.47. GA approach

4.10.1 GA Parameters

All notations used in the following correspond to the notations used for the MILP approach defined in sections 4.1 and 4.8. The parameters required by the genetic algorithm include

- the *number of generations* N,
- the *population size* p, defining the number of chromosomes in a population,
- the *population replacement size* r, specifying that for each generation a new population is created consisting of r chromosomes created by recombination and mutation, and $p - r$ chromosomes of the old population,
- the *selection type*, defining if only the fittest chromosomes are selected for new generations or also some others using probability values,
- the *crossover type* (one point, two point, etc.),

- the *crossover probability*,
- the *mutation type*,
- the *mutation probability*,
- a *stopping criterion*, e.g. a number q of generations without improvement of the best solution, etc.

This set of parameters is only a part of all possibilities to influence the optimization process. The complete set of parameters represents powerful possibilities to steer the genetic algorithm.

4.10.2 GA Encoding

As mentioned before, using the PGAPACK library allows to encode discrete values by genes which simplifies the encoding. Mapping and binding aspects are encoded on the chromosome. Scheduling is not encoded on the chromosome, but considered by the fitness function. For each *computation node* $v_i \in V^I$ of a partitioning graph G^P two genes are required, for each *communication node* $v_i \in V^{RI} \cup V^{WO} \cup V^W$ only one gene is required. A READ-node $v_i \in V^R$ is considered with the help of its predecessor WRITE-node as described in section 4.8.5. Thus, the total number of genes n_{genes} of the chromosome for encoding the hardware/software partitioning problem is defined by equation 4.63.

$$n_{genes} = 2* \mid V^I \mid + \mid V^{RI} \cup V^{WO} \cup V^W \mid \tag{4.63}$$

The structure of the chromosome representing the mapping of a partitioning graph $G = (V, E, C, I)$ to a target technology $\mathcal{T} = (\mathcal{V}, \mathcal{E})$ is defined as follows:

Genes $g_1, \ldots, g_{|V^I|}$**:** For each computation node $v_i \in V^I$ gene g_i determines on which target architecture component t_k node v_i is mapped. The value range for the alleles a_i of these genes is $\{0, \ldots, \mid \mathcal{PU} \mid -1\}$

Genes $g_{|V^I|+1}, \ldots, g_{2*|V^I|}$**:** For each computation node $v_i \in V^I$ gene $g_{i+|V^I|}$ is only required if allele a_i represents a hardware component h_k. Then, gene $g_{i+|V^I|}$ determines to which hardware instance $h_{j,l,k}$ of system component $c_l = I(v_i)$ node v_i is bound. Thus, the value range of $a_{i+|V^I|}$ is $\{0, \ldots, \mid instances\_of(c_l) \mid -1\}$.

Genes $g_{2*|V^I|+1}, \ldots, g_{n_{genes}}$**:** For each *communication node* $v_i \in V^{RW}$ of the partitioning graph only one gene defines which communication channel will implement the WRITE- or READ-routine. WRITE- and READ-nodes implementing a data transfer have to be mapped to the same communication channel obviously. Therefore, the communication channel required for the data transfer is encoded by the gene for the WRITE-node and the genes for the

READ-nodes are not necessary. The value range for a communication node v_i is encoded by $a_{i+2*|V^I|} = k \subseteq \{0, \ldots, |\mathcal{E}| -1\}$, including all communication channels b_k being able to implement the data transfer. Each of these genes is only considered if it really represents an interface. This depends on the mapping and binding of the computation nodes encoded by genes $g_1, \ldots, g_{2*|V^I|}$.

The encoding is illustrated by the well-known 4-band equalizer example in figure 4.48.

Example 18:

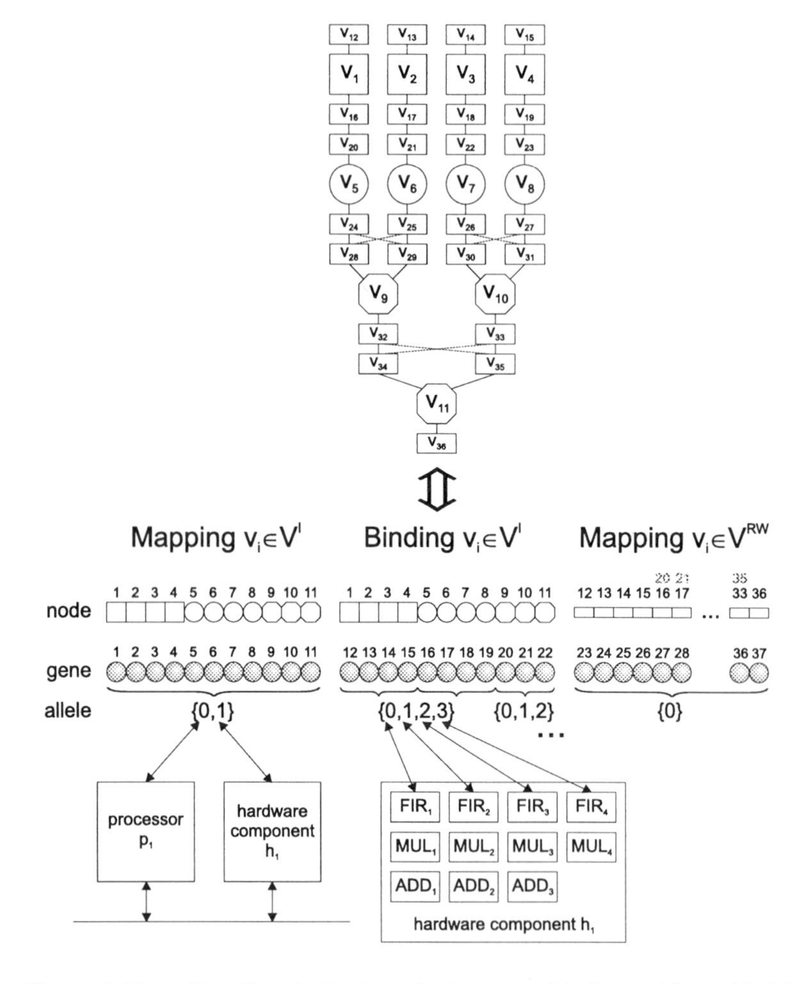

Figure 4.48. Encoding the hardware/software partitioning problem with GA

The mapping is determined by genes $g_1, \ldots, g_{11}$. Each computation node $v_i \in V^I$ may be mapped onto processor p_1 represented by allele 0 or hardware component h_1 represented by allele 1. Genes $g_{12}, \ldots, g_{22}$ encode the binding information. The most expensive hardware solution would contain four filters, four multipliers and three adders. Therefore, the alleles for the filters (encoded by $g_{12}, \ldots, g_{15}$) have a value range of $\{0, \ldots, 3\}$. Finally, genes $g_{23}, \ldots, g_{37}$ determine the binding of a communication node to a communication channel implementing R/W-accesses if necessary. In this example, only one communication channel b_1 is available, encoded by allele 0. The WRITE-node v_{16} READ-node v_{20} are encoded by the same gene, because they have to use the same communication channel.

REMARK:. Resource constraints and/or user-defined mapping and binding constraints can be very easily adapted by restricting the value ranges of the alleles. If, for example, v_1 should be mapped onto processor p_1, the value range of g_1 is limited to $\{0\}$. A generated solution represents a valid hardware/software implementation meeting all resource restrictions. However, timing constraints may be violated which cannot be prevented by restricting the value ranges of the alleles. Timing constraints will be affected by a violation metric included in the fitness function. It should be mentioned that this encoding needs no *repair mechanism*.

4.10.3 GA Fitness Function

During the evaluation phase, the *fitness function* F evaluates each *string* $s = \{a_1, \ldots, a_n\}$ encoded on a chromosome. All resource costs can be derived from eq. 4.19-4.21, introduced for the MILP approach, because of the following equations:

Let $G^P = (V, E, C, I)$ be a partitioning graph; $V = V^I \cup V^{RW}$ (def. 1).
Let $N = \mid V^I \mid$ be the number of computation nodes of the system.
Let a_i be the allele of the i-th gene on a chromosome.

$$\forall i \leq N : \forall p_k \in \mathcal{P} : \quad a_i = k \Rightarrow Y_{i,k} = 1 \tag{4.64}$$

$$a_i \neq k \Rightarrow Y_{i,k} = 0 \tag{4.65}$$

$$\forall i \leq N : \forall h_k \in \mathcal{H} : \quad a_i = k \Rightarrow X_{i,k} = 1 \tag{4.66}$$

$$a_i \neq k \Rightarrow X_{i,k} = 0 \tag{4.67}$$

$$\forall i \leq N : \forall h_k \in \mathcal{H} : \quad a_i = k \wedge a_{i+N} = j \Rightarrow x_{i,j,k} = 1 \tag{4.68}$$

$$a_i = k \wedge a_{i+N} \neq j \Rightarrow x_{i,j,k} = 0 \tag{4.69}$$

The fitness function evaluates a string $s = (a_1, \ldots, a_n)$ encoded on a chromosome by using equation 4.70.

$$F(s) = f_4 * C^a(s) + f_3 * C^{dm}(s) + f_2 * C^{pm}(s) + f_1 * C^t(s) + f_0 * C^{viol}(s) \tag{4.70}$$

The fitness function considers the resource costs including the total amount of hardware area C^a, data memory C^{dm} and program memory C^{pm}. In addition, the system execution time C^t, computed by a *scheduler*, is taken into account. Finally, a new metric C^{viol} for *constraint violation* is required.

These cost values are scaled by user-defined factors $f_0, \ldots, f_4$. Constant f_0 is very important to support hard constraints. In this case, it must be guaranteed that $f_0 \gg f_1, f_2, f_3, f_4$ to decrease the probability that chromosomes representing solutions violating some constraints are selected for the next population. With the help of constants f_1, f_2, f_3, f_4 the importance of minimizing hardware area, data/program memory usage or total execution time can be weighted. For example, the optimization goal *"hardware minimization under timing constraints"* can be implemented by using $f_0 \gg f_4 \gg f_1, f_2, f_3$. For each generated solution the genetic algorithm computes a fitness. Therefore, the fitness function has to compute the fitness value as fast as possible.

4.10.3.1 Resource Costs. The resource costs include hardware area and software data/program memory. They can be computed in linear time by using two matrices as depicted in figure 4.49: a 3-dimensional matrix MAT^{hw} for all nodes mapped to hardware components and a 2-dimensional matrix MAT^{sw} for all nodes mapped to processors.

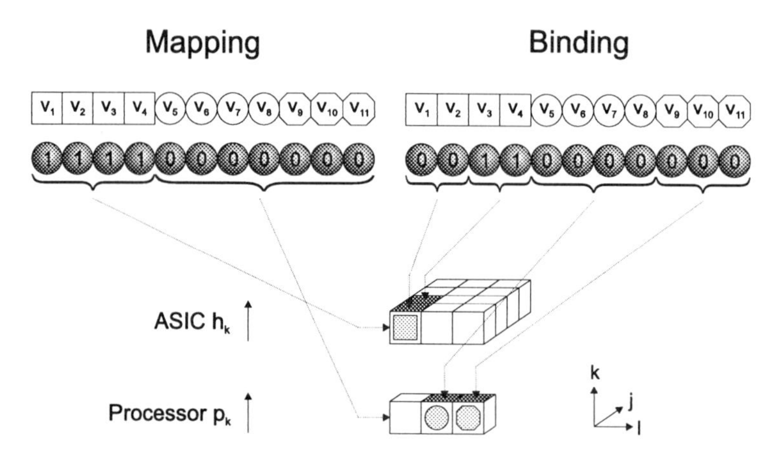

Figure 4.49. Matrices to compute resource costs in linear time

The axes of MAT^{hw} contain the different system components c_l (x-axis), the j-th hardware instance of c_l (y-axis), and the target architecture t_k (z-axis). The y-axis is missing for processors, because, as described before, concurrency of functions cannot be exploited on processors. To estimate the memory usage of these matrices, the maximum values of the axes should be determined. The number n_x of system components c_l on the x-axis is $n_x = \mid C \mid$. The number of instances of a system component on the y-axis is defined by n_y. An upper bound for n_y is defined by: $n_y \leq \mid V^I \mid - \mid C \mid +1$. The number of elements on the z-axis is either $|\mathcal{H}|$ for the hardware or $|\mathcal{P}|$ for the software matrix. Therefore, the total size S of matrix elements for both matrices can be computed by

$$S \leq (|C| * (|V| - |C| + 1) * |\mathcal{H}|) + (|C| * |\mathcal{P}|).$$

Using these matrices MAT^{hw} and MAT^{sw}, algorithm 3 computes the resource costs in time $O(|V^I|)$.

Algorithm 3: UpdateResourceCosts

```
(1)  algorithm UpdateResourceCosts;
(2)      input  mark, a_1, ... a_{2|V^I|}  : natural;
(3)      output C^{pm}, C^{dm}, C^a       : natural;
(4)  {
(5)      variable i, j, k, l      : natural;
(6)      variable MAT^{hw}        : array[|C|, |V| - |C| + 1, |H|] of natural;
(7)      variable MAT^{sw}        : array[|C|, |P|] of natural;
(8)      variable c_l             : SystemComponent;
(9)      variable p_k             : Processor;
(10)     variable h_k             : HardwareComponent;
(11)     variable t_k             : ProcessingUnit;
(12)     variable C_k^{pm}, C_k^{dm}, C_k^a : natural;
(13)
(14)     C^a = 0; C^{dm} = 0; C^{pm} = 0;
(15)     forall h_k ∈ H do { C_k^a = 0; }
(16)     forall p_k ∈ P do { C_k^{dm} = 0; C_k^{pm} = 0; }
(17)     forall v_i ∈ V^I do
(18)     {
(19)            k = a_i;
(20)            j = a_{i+|V^I|};
(21)            c_l = I(v_i);
(22)            if t_k ∈ P then
(23)            {
(24)              if MAT_{l,k}^{sw} ≠ mark then
(25)              {
(26)                C_k^{dm} = C_k^{dm} + c_{l,k}^{dm}; C_k^{pm} = C_k^{pm} + c_{l,k}^{pm};
(27)                C^{dm} = C^{dm} + c_{l,k}^{dm}; C^{pm} = C^{pm} + c_{l,k}^{dm};
(28)                MAT_{l,k}^{sw} = mark;
(29)              }
(30)            }
(31)            else
(32)            {
(33)              if MAT_{l,j,k}^{hw} ≠ mark then
(34)              {
(35)                C_k^a = C_k^a + c_{l,k}^a;
(36)                C^a = C^a + c_{l,k}^a;
(37)                MAT_{l,j,k}^{hw} = mark;
(38)              }
(39)            }
(40)     }
(41) }
```

First, all costs are initialized in lines 14 to 16. Then the hardware and software costs are iteratively accumulated (lines 17 to 40). If v_i is mapped to a processor (line 22) and the costs for implementing system component c_l as a function on processor t_k have not been incorporated (line 24), then the cost values of the processor and the global software costs are updated. Afterwards, the matrix element for implementing c_l on processor t_k is marked (line 28). Therefore, other instances of c_l which are mapped to t_k do not affect additional costs. The hardware costs are updated accordingly in lines 33 to 38.

4.10.3.2 Scheduling. After the genetic algorithm has determined a string s for the mapping and binding, the *scheduling* phase has to compute a schedule for this mapping. Here, the list scheduling approach described in section 4.9.3 is used to compute the total execution time for the colored partitioning graph. The time complexity of the algorithm to compute such a schedule is $O(|V|^2)$.

4.10.3.3 Violation. The violation metric is a very important metric, because strings s which represent illegal solutions have to be penalized. As a consequence, the additional costs of the violation metric lead to a reduced probability of s being selected for the population of the next generation. An important issue is that the degree of violation is measured. A string representing a solution with a small constraint violation will get lower additional costs than another solution with a hard violation. Otherwise, it would be difficult to find a solution if the solution space is very small.

Algorithm 4: Violation

```
(1)  algorithm Violation;
(2)      input G          : PartitioningGraph; f: real;
(3)      output C^viol    : real;
(4)  {
(5)      variable r^t, r_k^a, r_k^dm, r_k^pm, ...  : real;
(6)      variable p_k                               : Processor;
(7)      variable h_k                               : HardwareComponent;
(8)
(9)      r^t = Max(0, C^t - Max^t);
(10)     forall h_k ∈ H do
(11)         r_k^a = Max(0, C_k^a - Max_k^a);
(12)     forall p_k ∈ P do
(13)     {
(14)         r_k^dm = Max(0, C_k^dm - Max_k^dm);
(15)         r_k^pm = Max(0, C_k^pm - Max_k^pm);
(16)     }
(17)     ...
(18)     (19)                    C^viol = f * r^t;
(20)     forall h_k ∈ H do C^viol = C^viol + f * r_k^a;
(21)     forall p_k ∈ P do C^viol = C^viol + f * r_k^pm + f * r_k^dm;
(22)     ...
(23) }
```

In algorithm 4, only a part of the constraints is considered, illustrated by the dots in lines 17 and 22. First, all relative violation metrics are computed (lines 9 to 17). If a cost metric C^x violates the corresponding maximum constraint MAX^x, then $C^x - MAX^x > 0$. If the constraint is not violated, the relative violation metric is set to 0 by the maximum function (`Max(...)`). After the relative violation metrics have been computed, the absolute violation metric C^{viol} accumulates all violations multiplied by the user-defined violation factor f (lines 19 to 22).

4.10.4 Analysis of the GA Approach

To summarize, for each solution the genetic algorithm computes the fitness value. The resource costs are computed in computation time $O(|\ V^I\ |)$, the total execution time using list-scheduling in time $O(|\ V\ |^2)$ and the violation afterwards in time $O(|\ \mathcal{H}\ |+|\ \mathcal{P}\ |)$. Therefore, the computation time for the scheduling is the time-critical part. Hence, the time complexity to compute the fitness functions is $O(|\ V\ |^2)$. This run-time analysis shows that, in contrast to the MILP approach, genetic algorithms can also be applied to larger systems, because there is no exponential growth in computation time.

4.10.5 Results III: HW/SW Partitioning based on GA

In this section, the results of hardware/software partitioning based on genetic algorithms will be presented. Before this is done, the influence of the user-defined genetic parameters will be examined first.

4.10.5.1 Influence of Genetic Parameters. A large number of partitions has been computed for four different different systems (`eq7`, `al4`, `fz1`, `fz2`) for two different timing constraints, a very hard and an average one. For all these different examples the following genetic parameters have been tried out:

- the *crossover type* (either *uniform*, *one-point* or *two-point*),
- eight different values for the *mutation rate* (fixed or depending on the string length n),
- five different values for the *crossover rate*.

In total, the genetic algorithm has computed 960 different designs. Each of these runs has included 200 generations with a population size of 20 chromosomes. 15 chromosomes have been replaced per generation.

The following results can be summarized: First, the crossover type should be uniform, because it has been pointed out that, on the average, uniform crossover is superior to one-point or two-point crossover for this particular problem. In the following, the results are presented for uniform crossover. In figure 4.50, the average relative deviation from the best solution for a set of different mutation and crossover probabilities is depicted.

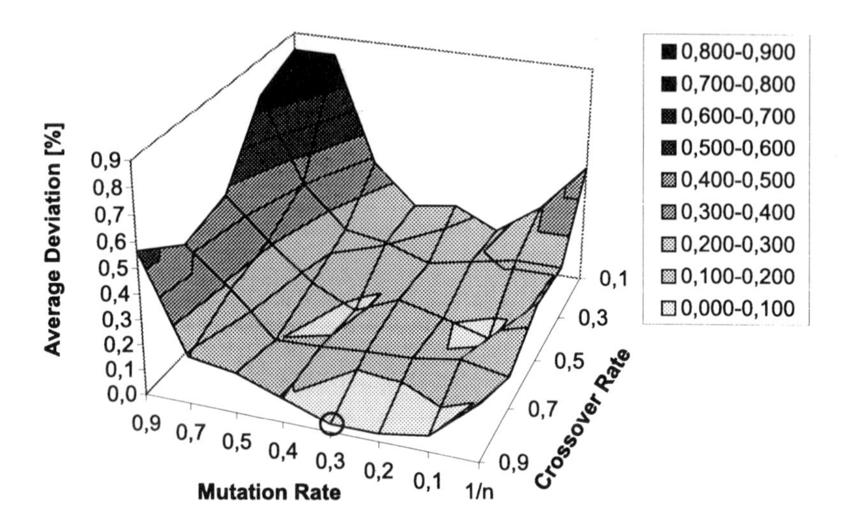

Figure 4.50. Average relative deviation to best solution

These values have been computed by scaling the results for each design of a system into a range of $0, \ldots, 1$ depending on the best and worst solution for this system. This scaling is required to guarantee that each design has the same influence on the total average. Then, the average for all these systems has been computed. The best solutions have been computed with a mutation rate of 0.3 and a crossover rate of 0.9. For these parameters the average relative deviation from the best solutions for all partitions is about 3%. To distinguish the influence of mutation and crossover rates, an average value has been computed for these rates using the results of figure 4.50. Figure 4.51 shows the results for different mutation and crossover rates separately.

Summarizing, it has been shown during benchmarking that the crossover rate has to be very high (about 90%) and the mutation rate has to be defined between 10 and 50%.

4.10.5.2 Partitioning Results.

The partitioning approach based on GA has been performed to compute 104 designs for 12 different systems. The parameters for the genetic algorithm have been defined as follows:

- uniform crossover with a crossover rate of 0.9,
- mutation rate : 0.3,
- 500 generations,

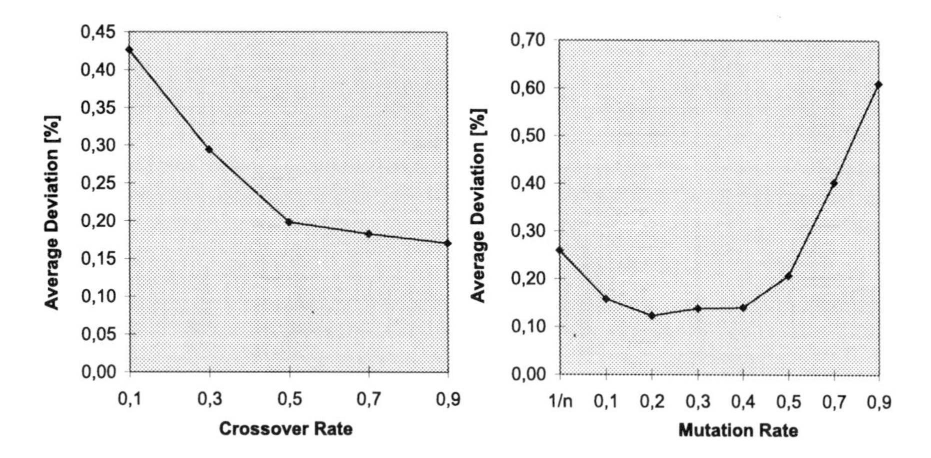

Figure 4.51. Influence of crossover and mutation rates

- population size of 20 chromosomes,
- for each generation 50% of all solutions are replaced by new solutions.

In figure 4.52, the resulting chip area of all designs is depicted.

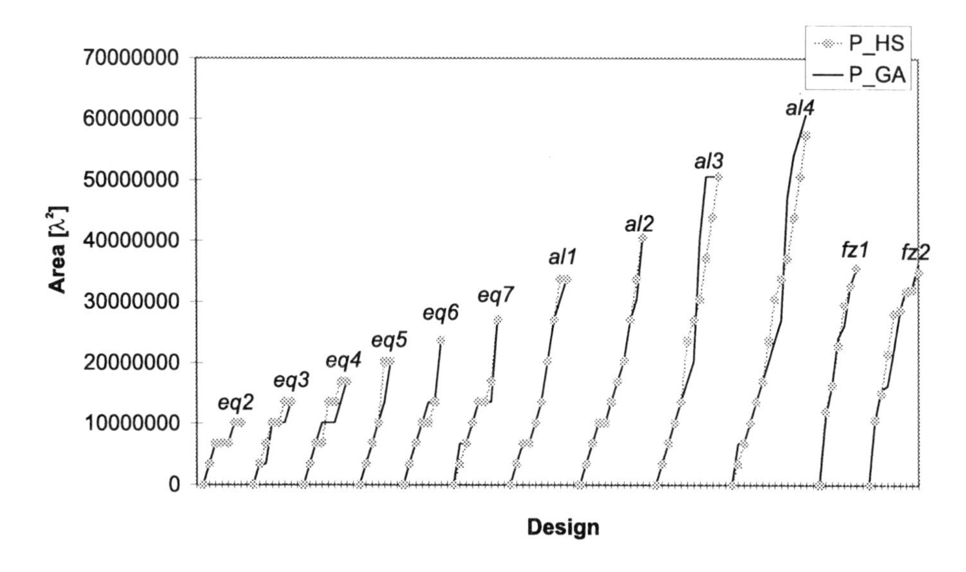

Figure 4.52. Chip area obtained by GA partitioning algorithm

Computed designs of both approaches are comparable. In figure 4.53, the deviation values of both approaches are presented. In 68 cases the algorithms have computed the same solution, in 21 cases the genetic algorithm has performed better, and in 15 cases the heuristic scheduling approach has computed better results.

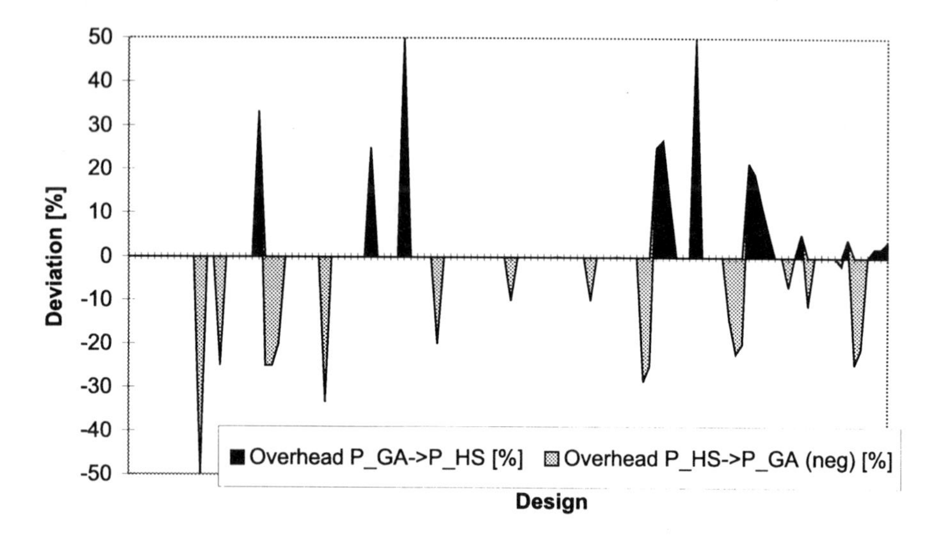

Figure 4.53. Deviation of computed hardware area for different partitioning algorithms

Accumulating the hardware area of all designs, the genetic algorithm finds partitions with 0.07% less chip area compared to the heuristic scheduling approach.

The comparison of the computation time is given in figure 4.54.

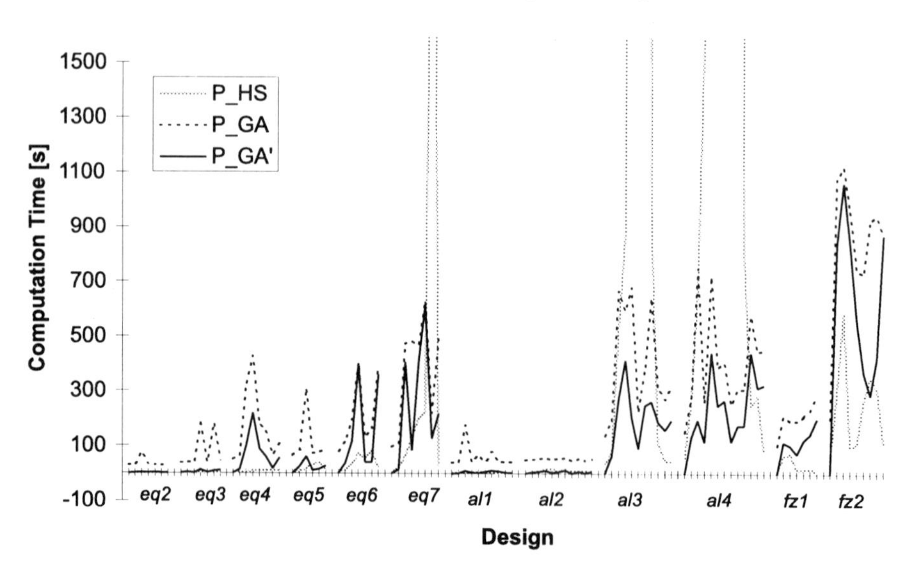

Figure 4.54. Computation time of partitioning algorithms

Curve P_GA represents the computation time for a genetic algorithm. Curve P_GA' contains the time points at which the best solution has been found by the genetic algorithm. Curve P_HS represents the computation time of the heuristic scheduling approach. Here, the computation time varies very strongly and therefore it is not predictable.

Table 4.7 summarizes all results obtained by this comparison concerning computation time.

Example	Computation Time [sec] average	worst-case
P_HS	1035	22368
P_GA	261	1120
P_GA'	141	1062

Table 4.7. Comparison of P_HS and P_GA for hardware/software partitioning

Clearly, the GA approach P_GA has been shown as superior compared to the heuristic scheduling approach concerning quality and computation time. For smaller examples, the heuristic scheduling approach P_HS may be better compared to GA, but with increasing complexity of the systems, the genetic algorithm approach will be advantageous. In addition to the results presented before, the GA approach has other advantages:

- It is well suited for *design space exploration*, because a set of solutions is computed, not only the optimal solution.
- GA is very stable. During benchmarking, in some cases problems occur when using MILP models. These problems were difficult to analyze and in some cases no result could be computed although solutions were existing (found by P_GA). These problems occured for different IP-solvers, like LP_SOLVE or OSL, but very often for different models. One reason for this is that some accuracy gets lost when real numbers of an MILP problem are written into a text file.
- P_GA can very easily be extended, e.g. the additional hardware costs which occur during the following co-synthesis step could also be considered during partitioning. This would be impossible when using P_MILP or P_HS, because most of these aspects cannot be modelled with linear terms.

Another extension to the partitioning problem will be described in the following section.

4.10.6 GA Approach for Extended HW/SW Partitioning

All algorithms presented before have computed hardware/software partitions under the assumption that for each node of the partitioning graph hardware and software costs are available. But, the hardware and software costs are restricted only to one solution. This restriction is very hard in particular on the hardware side, because there is a lot of freedom for alternative hardware implementations. The partitioning approach presented before takes into account the cheapest design for the fastest hardware implementation.

In contrast, the *extended partitioning problem* considers a set of different costs for alternative implementations. The extended partitioning approach in COOL, called P_GA_EXT in the following, allows to consider a complete AT-curve during the optimization step. The costs for different hardware implementations in COOL are computed during cost estimation (see section 4.4.2). But now additional control steps are offered to each basic block, leading to a reduced chip area. As a negative consequence of these additional control steps, the computation time explodes when solving the hardware synthesis problem by ILP in OSCAR. Therefore, another tool [Lore97] (based on genetic algorithms) has been developed for OSCAR solving these problems in short computation time while computing similar results.

The extended partitioning problem could also be formulated by MILP using an additional index for the mapping variables. But, the generated MILP models would not be solvable in acceptable time because of complexity reasons. In COOL, the extended partitioning problem is solved by using genetic algorithms. The encoding presented in section 4.10.2 is extended by using an additional gene for each computational node to encode the hardware implementation alternative as depicted in figure 4.55.

Example 19:

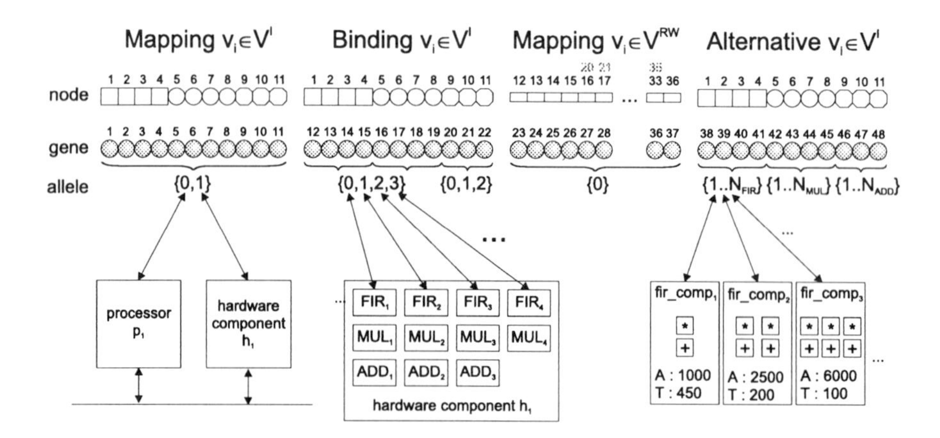

Figure 4.55. GA encoding for extended partitioning

In figure 4.55, the encoding of the extended partitioning problem for the 4-band equalizer is shown. In this example, N_{FIR} different hardware implementations for each FIR-filter $v_1, \ldots, v_4$ are considered during partitioning. The implementation alternatives for $v_1, \ldots, v_4$ are encoded by genes $g_{38}, \ldots, g_{48}$. The first implementation alternative fir_comp$_1$ is very cheap (chip area of 1000 λ^2) but very slow (450 ns execution time). This alternative is represented by allele 0. The third alternative is very fast (100 ns) but very expensive (6000 λ^2). The second one represents a compromise between performance and chip area.

The total number of required genes n_{genes} on the chromosome for encoding the extended partitioning problem is defined by

$$n_{genes} = 3* \mid V^I \mid + \mid V^{RI} \cup V^{WO} \cup V^W \mid \quad (4.71)$$

In addition to the genetic algorithm for the simple hardware/software partitioning problem, now the chromosomes have to be *repaired*, because it has to be ensured that nodes sharing the same hardware resources use the same hardware implementation alternative. The fitness function (see eq. 4.70) does not change, but the computation of the hardware costs now considers the selected hardware implementation alternative. The following example illustrates this fact:

Example 20:

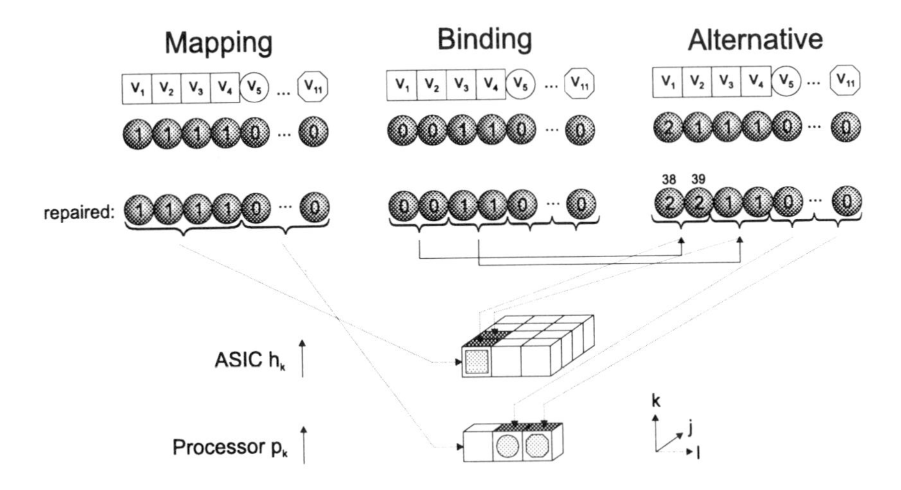

Figure 4.56. Hardware implementation alternatives in extended partitioning

In figure 4.56, a partition is depicted using two hardware instances of FIR-filters (a first one for v_1, v_2, a second one for v_3, v_4). The repair mechanism changes gene 39, because v_1 and v_2 share the same resource, but different alternatives are encoded. Therefore, the repair mechanism sets gene 39 to the allele of gene 38. The repaired chromosome represents a partition where the first hardware instance of an FIR-filter is implemented by the third alternative (encoded by allele 2) and the second FIR-filter is implemented by the second alternative (encoded by allele 1). The costs for the selected hardware implementation alternatives are inserted into the matrix.

Adapting these extensions to the original algorithm is very easy and will not be described in more detail. The advantages of considering a complete AT-curve instead a single hardware implementation are illustrated by the following example.

4.10.6.1 Results for Extended Partitioning based on GA.

The results obtained for extended partitioning approach will be described using system `al4` which has been introduced in section 4.9.4. In a first step, alternative hardware implementations have been computed for some components of `al4`. The results are depicted in figure 4.57.

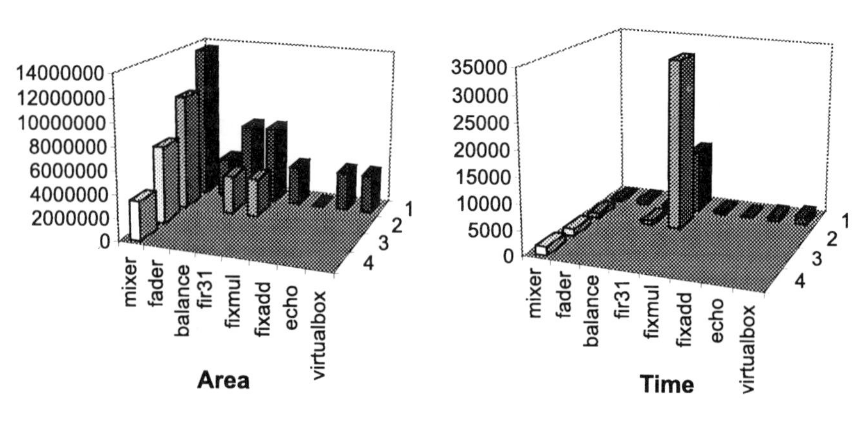

Figure 4.57. Alternative hardware implementations for components of `al4`

Four alternative hardware implementations have been computed for the `mixer` component, two alternatives for the `balance` ruler and also two alternatives for the `fir31` filter. For all other components a significant cost reduction could not be obtained. On the average, the fastest solutions are 130% faster than the slowest solutions, but additionally they require about 71% more chip area. These costs have been used to compute a set of partitions by three different approaches:

1. `P_GA_CHEAP`: The hardware/software partitioning problem `P_GA` has been solved considering the hardware costs for the cheapest implementations of all system components.

2. `P_GA_FAST`: The hardware/software partitioning problem `P_GA` has been solved considering the hardware costs for the fastest implementations of all system components.

3. `P_GA_EXT`: The extended hardware/software partitioning problem `P_GA_EXT` has been solved considering the hardware costs for all alternative hardware implementations of all system components.

In figure 4.58, the resulting AT-curves for all approaches are depicted.

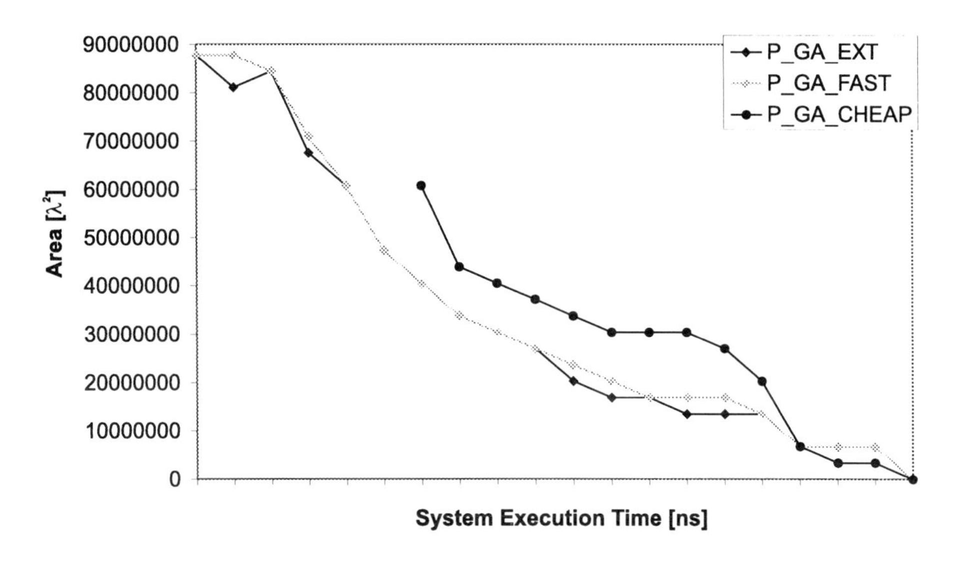

Figure 4.58. AT-curves of partitioning `al4` with different cost models

Clearly, `P_GA_CHEAP` suffers from the fact that no solutions can be found for hard timing constraints, because the fast implementation alternatives are expensive and therefore they are missing. In addition, the resulting partitions require on the average about 27% more chip area than `P_GA_EXT`. `P_GA_FAST` results in better partitions compared to `P_GA_CHEAP`. However, for soft timing constraints the computed partitions are often too expensive, because cheap implementation alternatives are not considered. On the average, `P_GA_FAST` requires about 9% more chip area compared to `P_GA_EXT`. `P_GA_EXT` requires 3 additional genes on the chromosome (for encoding the alternative hardware implementations of `mixer`, `balance` and `fir31`), but the computation times of all approaches are similar.

It can be summarized that `P_GA_EXT` is superior to `P_GA_CHEAP` and `P_GA_FAST`, because on the average it computes the better solutions compared to `P_GA_CHEAP` and `P_GA_EXT` in similar run-times.

5 HARDWARE/SOFTWARE CO-SYNTHESIS

After hardware/software partitioning has been executed, the partitioned system will be implemented. This design step is called *co-synthesis* and is illustrated in figure 5.1.

In a first *communication synthesis* step, the communication mechanism is determined by selecting a protocol and allocating memory. After this, the *specification refinement* step transforms the original implementation-independent system specification into a set of synthesizable hardware descriptions for the ASICs and a set of compilable software descriptions for the processors. The performed refinements correspond to the results of hardware/software partitioning and communication synthesis. The refined hardware and software specifications include the additional interface routines to exchange data. The abstract communication channels of the target architecture have been replaced by concrete communication protocols. The hardware descriptions are synthesized and the software specifications are compiled to implement the hardware/software system.

In the following, first the different tasks of co-synthesis will be introduced in section 5.1. Then, an overview of existing approaches in the area of co-synthesis is given in section 5.2. In section 5.3, the co-synthesis approach in COOL is presented. After this, the functionality of hardware/software systems generated by COOL is described in section 5.4. Then, the automation of the co-synthesis phase based on a state-transition graph is introduced in section 5.5. After the

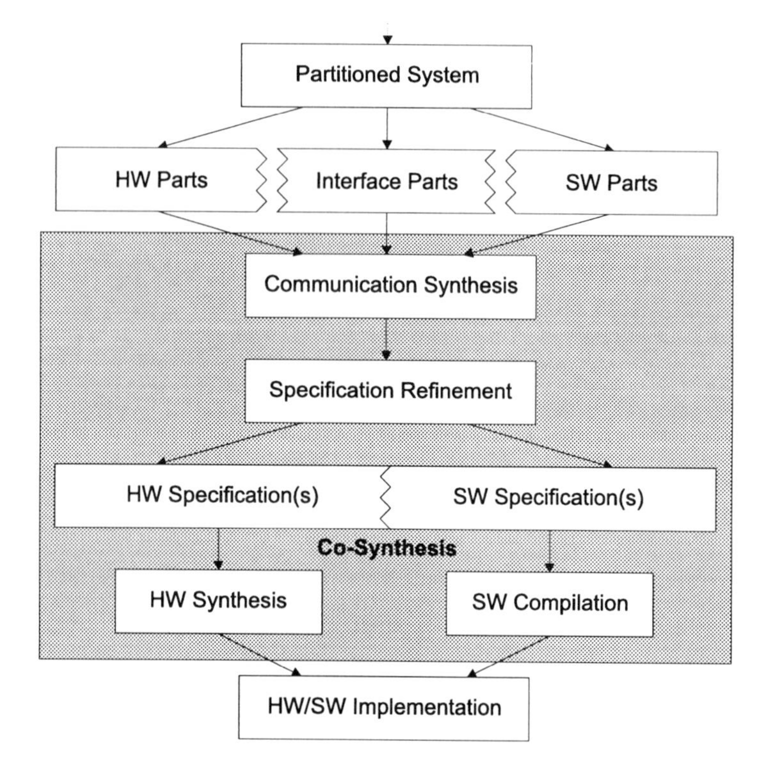

Figure 5.1. Overview of co-synthesis

refinement steps in COOL have been described in section 5.6, some results will be presented in section 5.7.

5.1 The Co-Synthesis Problem

Co-synthesis combines a lot of different tasks. As a consequence, a variety of different problem formulations can be found in the literature. Very often some notions are used in different meanings. Therefore, this section should introduce the notions related to co-synthesis used in this book. In figure 5.2, an overview of the most important tasks in co-synthesis is given.

The tasks of communication synthesis and specification refinement will be described next.

5.1.1 Communication Synthesis

Communication synthesis consists of selecting the protocols and the interfaces required by different subsystems for communication. *Protocol selection* allows

- Co-Synthesis
  - Communication Synthesis
    - Protocol Selection
    - Memory Allocation
  - Specification Refinement
    - Control-Related Refinement
      - Insertion of Synchronization Mechanisms
      - Generation of Runtime Environment
    - Data-Related Refinement
      - Variable Folding
      - Memory Address Translation
    - Interface Synthesis
      - Bus & Protocol Generation
      - Arbitration
      - Refinement of Incompatible Interfaces
  - Software Compilation & Hardware Synthesis

Figure 5.2. Tasks of co-synthesis

the designer to explore the design space, for example, by changing the protocols, the bus topology or the memory sizes. In *memory allocation*, for each variable of the system specification one or more memory cells are allocated in local or global memory. The number of cells to be allocated depends on the bit width of the variable and the memory width. The following example will illustrate the task of communication synthesis.

Example 21:

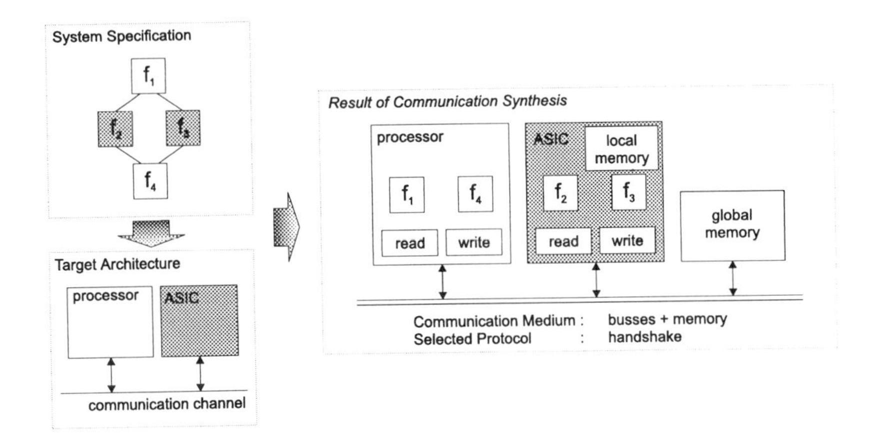

Figure 5.3. Communication synthesis

The original system specification, consisting of four functions $f_1, \ldots, f_4$, has been partitioned onto a single-processor-single-ASIC target architecture. Functions f_1 and f_4 have been mapped to the processor, f_2 and f_3 to the ASIC. The communication synthesis step (see figure 5.3) determines that communication should be based on shared memory communication. Thus, the abstract communication channel is replaced by a network of busses and a global memory. The protocol should be a simple handshake protocol, being able to access data with pre-defined `read`- and `write`-routines. Finally, local memory is added to the ASIC for storing internal data.

5.1.2 Specification Refinement

Specification refinement represents the design step transforming the implementation-independent system specification into a mixture of functional and structural components by adding implementation details. The *refined specification* reflects the results of

- hardware/software partitioning and
- communication synthesis.

The functionality of the original system specification and the refined specification has to be equivalent. The following example, depicted in figure 5.4, will illustrate the specification refinement step.

Example 22:

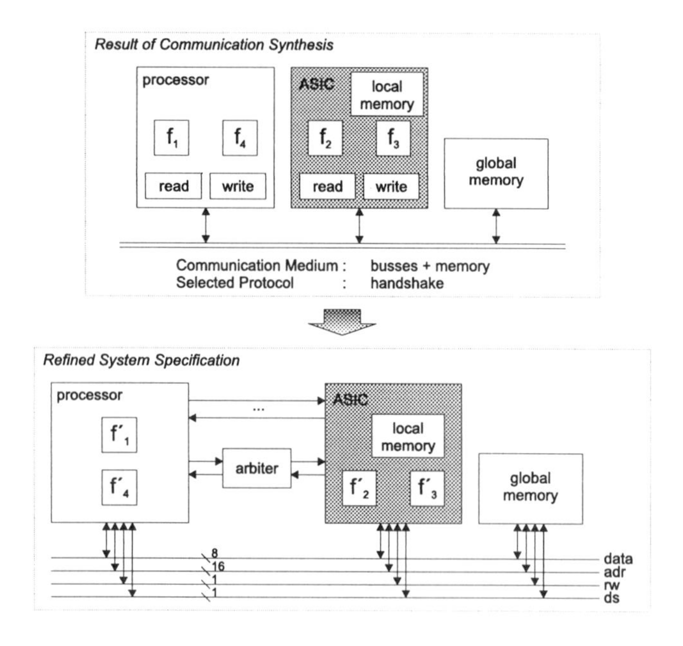

Figure 5.4. Specification refinement

Functions $f_1, \ldots, f_4$ of the original system specification have been refined into $f'_1, \ldots, f'_4$ during specification refinement using, for example, the `read`- and `write`- routines of the chosen protocol. In addition, the sizes of the busses have been computed and additional control signals between processor and ASIC have been generated. Finally, an arbiter has been instantiated to prevent bus conflicts.

Specification refinement is very important for the following reasons:

1. The error-prone task of inserting interfaces and the corresponding protocols between the target architecture components is automated.

2. The automation of this design step enables the designer to compare different implementation alternatives in short computation time allowing *design space exploration.*

3. The refined system specification represents a complete description of the system including all implementation and communication issues. Therefore, it can serve as an input to

   - *simulation* tools to validate the correct functionality,

   - *functional verification* tools to verify the correct functionality,

   - *hardware synthesis* and software *compilation* tools to implement the system.

Subtasks of specification refinement can be classified into three different categories (see figure 5.2):

- control-related refinement,

- data-related refinement,

- interface synthesis.

It should be mentioned that the execution order of these refinement steps is not fixed. In the following, these steps will be described in more detail.

5.1.2.1 Control-Related Refinement.

Control-related refinement steps represent the insertion of *synchronization* mechanisms corresponding to the execution sequence computed during hardware/software partitioning.

Insertion of Synchronization Mechanisms

If two components have been mapped to different processing units and if they have been sequentially scheduled, additional control signals are required for synchronization. There are different *synchronization mechanisms* to implement this sequential execution (see figure 5.5):

Synchronization by message passing: On the hardware side, the ASIC can be synchronized by *message passing* which implements a direct communication. In figure 5.5A, an example is given. The ASIC is *blocked* until the processor finishes the execution of function $\mathtt{f}_1$ and sets the `start`-signal to `1`. The message passing synchronization can also be applied to processors if some pins of the processor are available for sending/receiving messages.

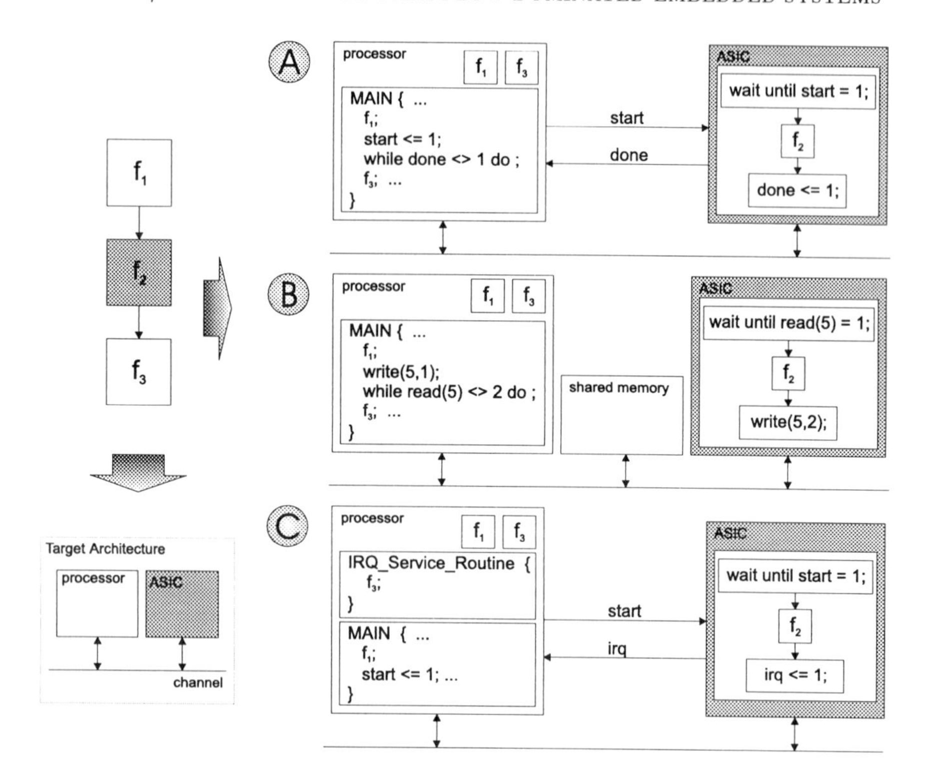

Figure 5.5. Control-related refinement

Synchronization by shared memory: If no unused pins are available for message passing, another possibility is synchronization using a *shared memory*. In such a case, certain memory cells are repetitively accessed until a certain value is read. This method may be very inefficient, because clock cycles are wasted. In figure 5.5B, an example using memory cell `memory[5]` is given.

Synchronization by interrupt: A better solution for processors is the usage of *interrupts*, as depicted in figure 5.5C. In such a case, the specification has to be refined by generating an *interrupt service routine*. In the example given, the ASIC sets the `irq`-signal to `1`. Thus, the processor starts its interrupt service routine executing function $\mathtt{f}_3$.

Runtime Environment

If the complete hardware/software system is controlled by hardware, an additional component has to be generated steering the target architecture components statically or dynamically corresponding to the computed schedule. This component is called a *run-time scheduler*.

5.1.2.2 Data-Related Refinement.

Data-related refinement steps transform accesses to data (variables) used in the implementation-independent system specification into memory accesses using the selected protocol. The following tasks for refining data accesses into the required communication mechanisms are included:

Data Type Conversion

Data type conversion has to be performed during hardware and software refinement. The main task is to transform the data type of variables of the implementation-independent specification into target-specific data types. On the hardware side, for example, only bit vectors are available. Therefore, hardware refinement has to replace all variable declarations by bit vector declarations.

Operation Replacement

Operation replacement is a task related to data type conversion, because now operations have to be transformed into target-specific equivalents. For example, a fixed-point multiplication has to be transformed into a sequential execution of integer multiplication and right-shifting if no fixed-point operation is defined for the selected processor or ASIC.

Variable Folding

Variable folding determines the assignment of each bit in a variable to a bit in a memory word. If the size of the variable is smaller than the memory width, then the variables can be assigned, for example, to the lower bits. If the bit width of a variable is greater than the memory width, then the words have to be stored in different memory cells, for example, in a sequence of cells. This sequence can be written using *little endian* or *big endian* assignment [Cohe81]. Little endian assigns the least significant bits (LSB) to the lowest address. In contrast, big endian assigns the most significant bits (MSB) to the lowest address. The difference between little endian and big endian is depicted in figure 5.6 showing a 64-bit value stored in a 16-bit width memory.

	adr	*adr+1*	*adr+2*	*adr+3*
little endian	data(15,...,0)	data(23,...,16)	data(39,...,24)	data(63,...,40)
big endian	data(63,...,40)	data(39,...,24)	data(23,...,16)	data(15,...,0)

Figure 5.6. Little and big endian

Memory Address Translation

For each variable of the system specification one or more memory cells have been allocated during the memory allocation step. *Memory address translation* transforms all references to the variables into memory accesses to the allocated cells. Corresponding to the selected communication medium and protocol, the accesses to external data have to be refined as illustrated in figure 5.7.

Example 23:

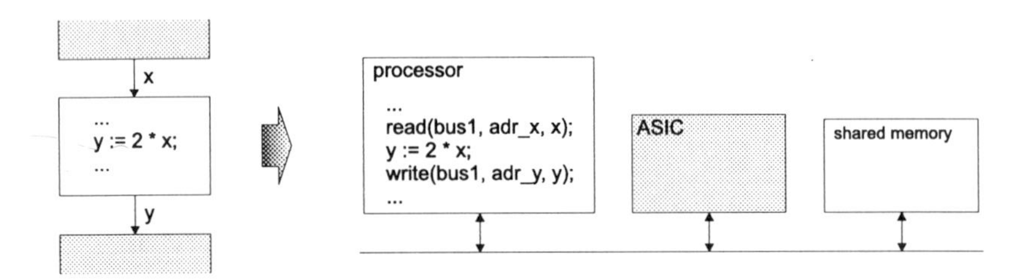

Figure 5.7. Refinement of accesses to external data

The value **x** which is required to compute **y** is computed on another processing unit. In this example, the abstract communication channel is refined by bus **bus1** connecting the processor and the ASIC with external memory for storing **x** and **y**. The refined specification works as follows: **x** is read from external memory from address **adr_x** using the bus. Afterwards, **y** is computed and written into external memory at address **adr_y**.

5.1.2.3 Interface Synthesis. *Interface synthesis* deals with implementation problems and can be defined as the realization of communication between components via both hardware and software elements. Before the tasks of interface synthesis are described, different communication models, communication channels and communication protocols will be introduced.

Communication Models.

Communication models can be divided into message passing and shared memory communication models. *Message passing communication* represents direct communication between sending and receiving processing units. Message passing communication is called *blocking* if the sending processing unit is waiting until the receiving processing unit is ready for the data transfer. In *non-blocking communication* sending and receiving processing units do not have to be synchronized. Thus, additional storage elements, e.g. FIFO buffers, are required to prevent the loss of data. *Shared memory communication* uses a shared storage medium which can be accessed by multiple processing units. Sending and receiving processing units do not have to be synchronized.

Communication Channels.

The *communication channel* may be implemented, for example, by dedicated lines, a single bus (without memory), a bus combined with memory for storing data, or FIFO buffers (see figure 5.8).

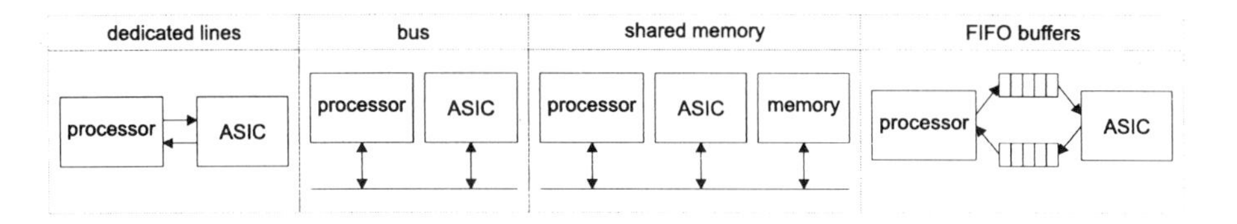

Figure 5.8. Communication channels

Dedicated lines: *Dedicated lines* are used to wire different processing units. Very often, the processing units have local memory and exchange data using these wires. Dedicated lines implement a *point-to-point connection* between the sending and receiving process which communicate using a blocking message passing protocol.

Bus: A *bus* represents a *multi-way channel* which allows to exchange data between several processing units. A sending processing unit writes a message onto the bus, using a blocking message passing protocol, and a set of receiving processing units may read this message. Thus, using a bus as the communication channel allows to implement a *broadcast mechanism*.

FIFO buffers: *FIFO buffers* use a non-blocking message passing protocol. Therefore, the sending processing unit sends messages without being synchronized with the receiving processing unit. The advantage is that no clock cycles are wasted for synchronization for the price of additional storage elements.

Shared memory: Communication based on *shared memories* also represents a non-blocking communication. The sending processing unit writes some data into the memory and the reading processing unit will read this data (if required) at a later point of time. The difference between communication based on FIFO buffers and shared memory is that FIFO buffers only allow a point-to-point connection whereas a shared memory uses a bus and therefore can be accessed by multiple processing units.

Table 5.1 summarizes the characteristics of the different types of communication channels.

Comm. Channel	Comm. Model	(non)-blocking	Topology
dedicated lines	message passing	blocking	point-to-point
bus (without memory)	message passing	blocking	multi-way
FIFO	message passing	non-blocking	point-to-point
shared memory	shared memory	non-blocking	multi-way

Table 5.1. Communication channels

Communication Protocols.

The *communication protocol* may be a simple *handshake protocol*, such as *2-phase* or *4-phase handshake* or a more complex protocol, such as RS-232, CAN[1], USB[2], PCI[3] and others. An overview of communication protocols used for embedded systems is given in [UpKo94].

Tasks of Interface Synthesis.

After the different communication models, channels and protocols have been introduced, now the tasks of interface synthesis will be described. The following subtasks are performed during interface synthesis to realize the communication between ASICs and processors.

Bus and Protocol Generation

Bus generation tries to compute the bus width resulting in minimum implementation costs while satisfying all performance constraints, such as data transfer rates. In figure 5.4, an example of this step is given. The computed bus widths are 16 bit for the address bus and 8 bit for the data bus.

Protocol generation refines the target architecture by inserting the required wires for the data bus, the address bus and the control signals. In addition, the `read`- and `write`-routines of the protocol have to be defined corresponding to the computed bit widths.

Arbitration

Bus arbiters are used to resolve bus access conflicts between different processing units connected to the same bus. This can be done using a fixed-priority or a dynamic-priority *arbitration scheme*. The *fixed-priority arbitration scheme* assigns the bus according to a statically determined priority. In contrast, the *dynamic-priority arbitration scheme* uses a priority determined at run-time, e.g. first-come-first-served.

Refinement of Incompatible Interfaces

If two processing units exchange data and each of them has a fixed protocol, then these protocols have to be made compatible. This task is called *protocol conversion*. In most cases, some additional *glue logic* is generated to make the protocols compatible.

---

[1] CAN: Controller Area Network
[2] USB: Universal Serial Bus
[3] PCI: Peripheral Component Interconnect

5.2 Related Work

CHINOOK [BCO95, COB95a], developed by Boriello et al. handles both communication synthesis and a wide range of interface synthesis problems. The following aspects of interface synthesis are supported: device drivers can be synthesized directly from timing diagrams [WaBo94]. *I/O port allocation* can be performed by two alternatives: for processors with I/O ports customized access routines reflecting the pin assignment [COB92] are generated. For processors without I/O ports, interfaces are implemented using *memory mapped I/O*. In this case, devices are allocated portions of the address space of the processor controlling them. CHINOOK allocates address spaces and generates the required bus logic and instructions [COB95b]. Communication synthesis, presented in [OrBo97], is based on a communication model which allows an easy retargeting to different protocols and architectures. The designer is able to map a high-level specification to an arbitrary but fixed architecture that uses particular bus protocols for interprocess communication. The communication synthesis tool has been fully integrated with a co-simulator, enabling the designer to get performance data to evaluate a given mapping.

SYMPHONY [LVM96a, LVM96b, VLM96], being part of COWARE, solves the co-synthesis and integration problem using a combination of architectural strategies, parameterized libraries and CAD tools. The approach is based on a simple communication protocol (called *synchronous wait protocol*) for all processing units. This protocol ensures that sender and receiver are synchronized with each other before the data transfer takes place. *Architectural templates* are built around processor cores in such a way that a processor can be integrated into the target architecture in the same way as any other hardware component. The architecture template contains the processor core, internal memory for storing program instructions and data, and an I/O unit implementing the hardware communication interface to the external environment. All these components are connected by the processor bus. SYMPHONY automates the design of the I/O unit which has to convert the communication protocol into the processor specific bus protocol (see [LiVe94]).

In COSMOS [IsJe95], communication synthesis is stated as an *allocation* problem [DIJ95, DMIJ97] allowing a wide design space exploration. This communication synthesis approach performs protocol selection and interface synthesis automatically using a communication library. The supported protocols in this library are low-level protocols, such as bidirectional handshake, single and dual FIFO.

The interface refinement step integrated in SPECSYN consists of two steps: *bus generation* and *protocol generation*. The bus generation algorithm [NaGa94a] determines the bit width of a bus implementing a set of abstract communication channels. Using the computed bus width, the protocol generation step [NaGa94b] defines the exact mechanism of transferring data over the bus. Different low-level protocols, such as full-handshake, half-handshake, fixed-delay and even hardwired ports are supported by SpecSyn. In addition, Narayan [NaGa95] presents an approach interfacing two fixed, incompatible protocols by generating hardware implementing an interface between them.

Vahid presents an object-oriented communication library (OOCL) [VaTa97] providing pre-implemented channel-based send/receive communication primitives (in C and VHDL) for a set of common protocols and components. The great advantage of this approach is that implementation becomes very easy, because OOCL is a library and therefore no synthesis tools are required to generate the communication behaviors.

Gong introduces an approach to interface synthesis based on model refinement [GGB96, GGB97]. This approach is able to transform the specification into a parameterizable implementation model. Four implementation models have been integrated and each of them is a mixture of local/global memories and busses. After the designer has selected the implementation model, the algorithm inserts the details of the interface.

Mooney presents an approach [MSM97] for synthesizing a *run-time scheduler*. This run-time scheduler represents a dynamic real-time scheduler in hardware and software. The overall control flow of the scheduler is implemented in hardware and the software contains the code for calling tasks.

Mickey performs rule based refinement [BMM98] of the control and data flow graph by inserting additional functional elements for the purpose of interfacing. The different categories of refinement include decomposing the event based and conditional transitions and incorporation of device drivers. Communication through both interrupt and polling has been considered.

5.3 Co-Synthesis in Cool

The goal of the co-synthesis approach realized in Cool is to refine the system specification by generating hardware specifications for synthesis and simulation in VHDL and software specifications for compilation in C. In figure 5.9, the design flow of the co-synthesis phase in Cool is depicted.

In a first step, a *state-transition graph* is generated representing the fundamental data structure during co-synthesis. This graph is computed using the colored partitioning graph. Then, the original system specification is refined into a set of specifications for the hardware and software parts. In this step, communication mechanisms for communication based on *shared memory* and *message passing* are inserted to replace the abstract communication channels. The refined hardware specifications represent the computational-intensive parts of the ASICs, e.g. an FIR-filter defined in the system specification. Thus, a refined hardware specification will be called *data path* in the following. To implement a complete hardware/software system, additional components are generated: a *system controller* steering the complete system, *data path controllers* to support

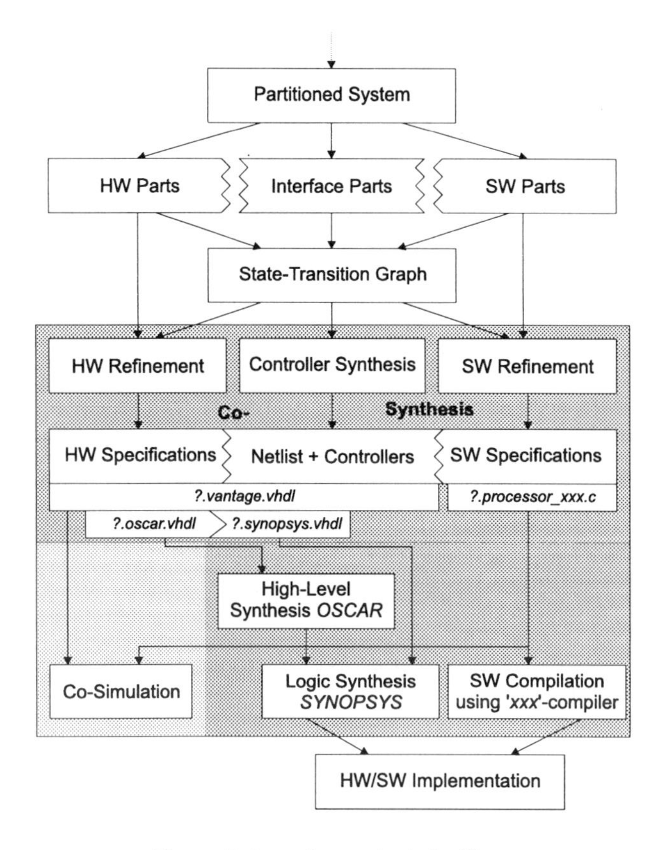

Figure 5.9. Co-synthesis in COOL

hardware sharing of the data paths, an *I/O controller* to communicate with the environment and a *bus arbiter* for each bus to prevent bus conflicts. The interfaces of processors are often very different. To describe a general control mechanism for hardware/software systems using a processor as the master of the system becomes difficult. Hardware allows the implementation of concurrent processes and therefore in COOL the additional pieces will be implemented in hardware. This general control mechanism for hardware/software systems will be described in section 5.4.

COOL not only generates the descriptions of the hardware components and the software running on the processors, but also a netlist wiring all these hardware components and processors. This netlist represents a complete description of the hardware/software implementation which can be co-simulated or implemented by hardware synthesis and software compilation. Simulatable VHDL descriptions are often not synthesizable and vice versa. Therefore, COOL generates different hardware specifications for each component. The simulation model is executed by the commercial VHDL simulator VANTAGE OPTIUM [Vant94]. Hardware synthesis is performed by the high-level synthesis tool OS-

CAR and by the commercial logic synthesis tool SYNOPSYS [Syno92]. There are two reasons to integrate two different synthesis tools:

1. OSCAR has two restrictions, leading to unsolvable problems when connecting designs synthesized with OSCAR to busses. First, OSCAR does not support *bi-directional signals*. Second, the *Z-value*, being an element of data type `std_logic` in VHDL, is necessary for specifying the behavior of busses. However, OSCAR only supports 0/1-values.

2. OSCAR is a high-level synthesis tool, very well suited for data-dominated VHDL specifications. It optimizes the costs for large basic blocks, but for control-dominated specifications it results in too expensive designs, because the average size of the basic blocks is very small. For these control-dominated VHDL specifications SYNOPSYS is used resulting in a drastic cost reduction compared to OSCAR.

For this reason, OSCAR and its backend tool OSBACK are used for synthesizing the refined data paths in COOL. Finally, the results of OSBACK and all additional hardware parts instantiated in the generated netlist are synthesized by SYNOPSYS.

With this separation, COOL was able to implement a Fuzzy-controller (specified with COOL) on a prototyping board containing a MOTOROLA DSP56001 [Moto94] and two Xilinx FPGAs 4005 [PG293b]. Without using SYNOPSYS in addition to OSCAR, this would have been impossible, because the available hardware area of these FPGAs is very small. This application study will be described in section 5.7.

5.4 Hardware/Software Systems generated by COOL

The functioning of hardware/software systems generated by COOL will be described with the help of the well-known equalizer example depicted in figure 5.10. To simplify the picture, some of the wires connecting components have been collapsed. These are represented by double lines.

The dotted components are additional components generated during the controller and netlist synthesis step shown in figure 5.9. For the following components synthesizable and simulatable VHDL descriptions are generated:

- system controller,
- I/O controller,
- bus arbiter,
- hardware data paths,
- hardware data path controllers and
- bus drivers.

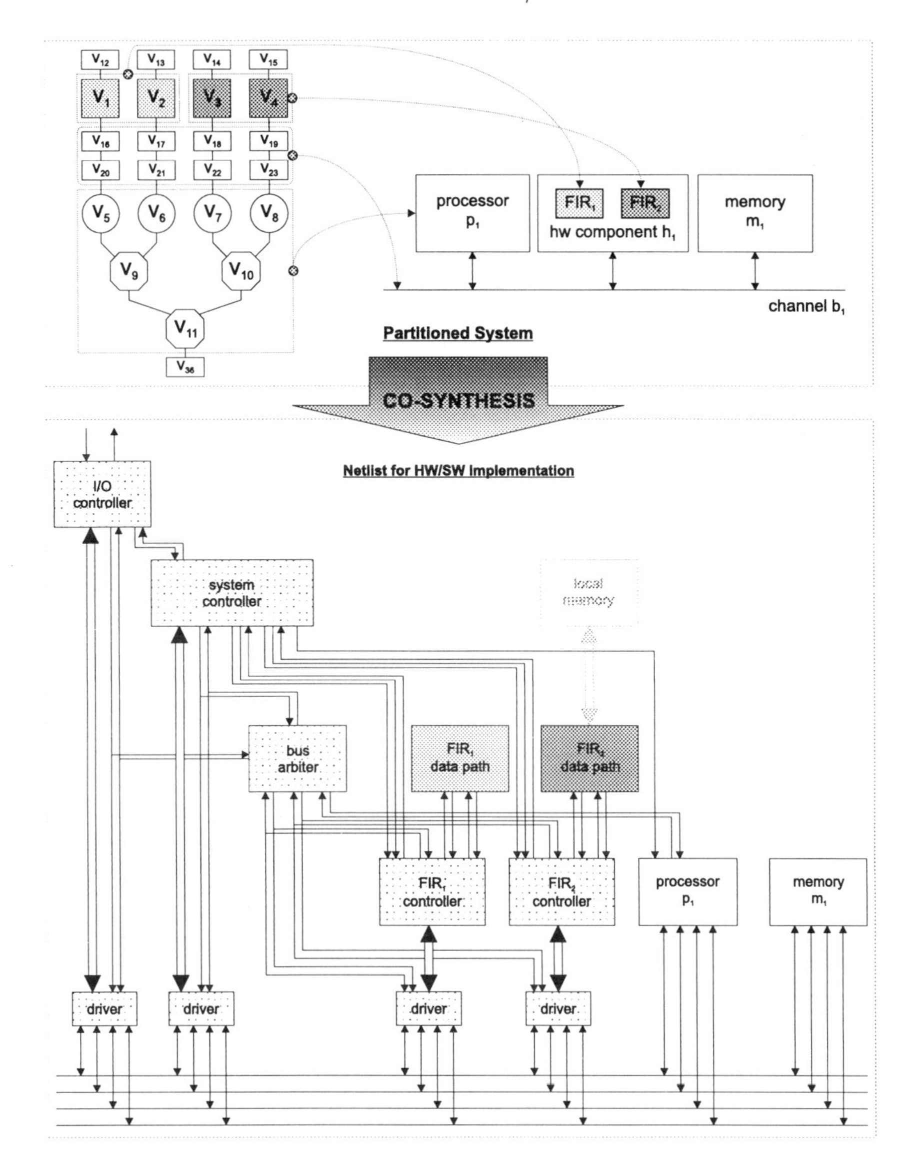

Figure 5.10. Hardware/software system generated by COOL

Processors and memories are fixed elements of the target architecture in COOL. Therefore, only pre-defined simulatable VHDL descriptions for

- processors,
- memories, and
- local memories.

are used which will be described in section 5.4.6.

In the following, the different components instantiated in the generated netlist will be described.

5.4.1 System Controller

The *system controller* is the heart of the complete hardware/software system. It has to be able

- to reset all processing units,
- to start the execution of a special function on each processing unit and
- to recognize the end of computation on each processing unit.

The system controller steers all processors and ASICs by activating them corresponding to the computed schedule during partitioning. Therefore, the system controller represents a *run-time scheduler*. There are different synchronization schemes integrated in COOL implementing this *start/stop mechanism*:

I. SW synchronization by interrupt and shared memory:
The processor has to execute different functions. In the example given (see figure 5.10), it computes multiplications and additions for nodes $v_5, \ldots, v_{11}$ by calling the corresponding function. The task of the system controller is to activate the processor to execute a certain function. Processors often offer the possibility to be interrupted from the environment and to read and write data from external memory. With these abilities, the following synchronization mechanism depicted in figure 5.11 can be implemented. Th synchronization mechanism can be applied in general to all processors ha interrupts and a memory interface.

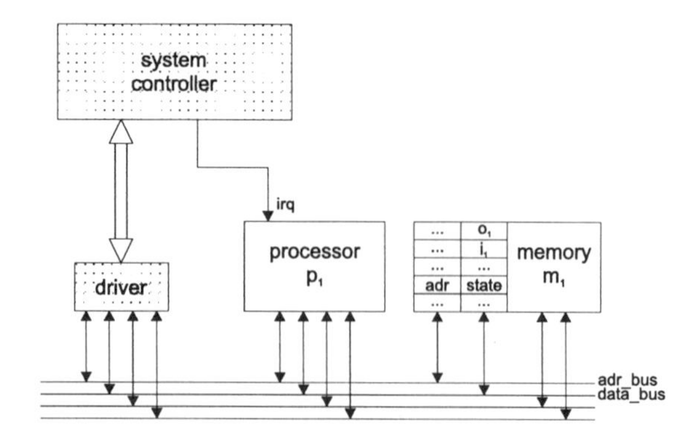

Figure 5.11. Synchronization scheme I for processors

The system controller and the processor communicate via shared memory. For a function implemented on the processor the following sequence is executed:

1. The system controller writes a number `state` into memory at address `adr`.
2. Then, processor p_1 is activated by setting the interrupt signal to `irq='0'`.
3. Processor p_1 reads `state` from external memory at address `adr`.
4. According to the value of `state`, the corresponding function f is executed by a) reading input data $\mathtt{i}_1$ from external memory, b) executing function f and c) writing computed result $\mathtt{o}_1$ into external memory.
5. After this sequence has been executed, processor p_1 writes `done_state` (an updated value of `state`) into external memory at address `adr`, identifying the end of computation.
6. The system controller uses *bus snooping* to read the `done_state` from the `data_bus` when the `address bus` is equal to `adr`.
7. Finally, the system controller updates the `state`-value of p_1.

It has to be mentioned that an incoming reset is encoded by a certain value of `state`. Summarizing, one `irq`-signal, one memory cell, two `WRITE`-accesses and one `READ`-access[4] are necessary to implement this synchronization.

II. SW synchronization by interrupt and message passing:
An alternative implementation for processors (see figure 5.12) is based on the idea of updating the state of the processor by the processor itself. Therefore, the processor has to implement some parts of the run-time scheduler to update its own state corresponding to the computed schedule.

In such a case, two additional unused pins are required. The first pin is used to distinguish an incoming `reset` and a normal activation (for executing a certain function) indicated by the incoming interrupt `irq`. The second pin `done` is required to indicate the end of computation of a certain function to the system controller. Thus, the memory accesses for exchanging the `state` of the processor can be eliminated, as illustrated by the following sequence:

1. Processor p_1 is activated by using the interrupt signal `irq`.
2. According to the internal value of `state`, the corresponding function is executed.
3. After this sequence has been executed, pin `done` is set to `done='1'`.
4. Variable `state` is updated.

---

[4]Reading the `done_state` in step 6 is realized by bus snooping which does not consume additional clock cycles!

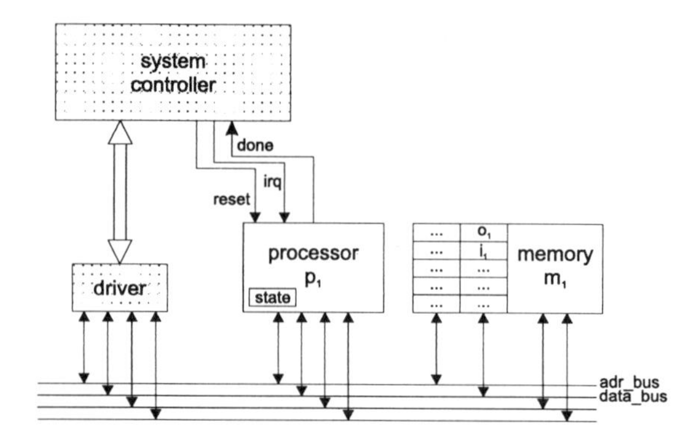

Figure 5.12. Synchronization scheme II for processors

Summarizing, one `irq`-signal and two additional pins are required to implement this synchronization mechanism.

III. HW synchronization by message passing and shared memory: The first synchronization mechanism presented for processors (shown in figure 5.11) can be adapted to hardware instances $h_{j,l,k}$ implementing components c_l on ASIC h_k. In such a case, a special `start`-signal is used instead of the interrupt signal `irq` in figure 5.11. The data path controller has to be sensitive to this `start`-signal. However, there are better alternatives (see IV. and V.) eliminating the memory accesses and thus optimizing the synchronization time.

IV. HW synchronization by message passing using start-signal: The second synchronization mechanism presented for processors (see figure 5.12) can also be adapted to hardware instances $h_{j,l,k}$ by using a `start`-signal instead of an interrupt. In this case, the data path controller has to implement some parts of the run-time scheduler to update its own state corresponding to the computed schedule. The advantage of this synchronization mechanism is that no memory accesses are required to update the state of hardware instance $h_{j,l,k}$.

V. HW synchronization by message passing using state-signal: The last synchronization mechanism for hardware instances also eliminates all `WRITE`- and `READ`-accesses required for synchronization. But in contrast to synchronization scheme IV., the state of $h_{j,l,k}$ is updated by the system controller instead of the data path controller. This is done by using a `state`-signal instead of the `start`-signal.

The synchronization (see figure 5.13) works as follows: For a function f implemented on hardware instance $h_{j,l,k}$, the following sequence is executed:

1. The system controller sets the `state`-signal to a certain value defining which function f has to be executed on hardware instance $h_{j,l,k}$.

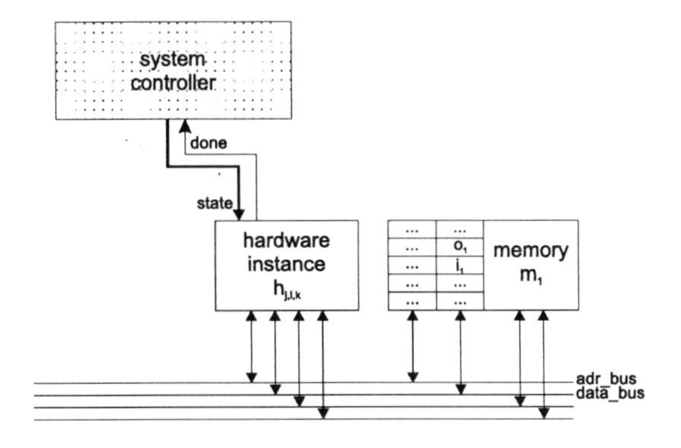

Figure 5.13. Synchronization scheme V for ASICs

2. Hardware instance $h_{j,l,k}$ starts execution of f, because `state` $\neq 0$.
3. When f has been executed, the `done`-signal is set.
4. The system controller recognizes that $h_{j,l,k}$ has finished, because `done='1'`. After this, the `state`-value of $h_{j,l,k}$ is updated.

This alternative is faster than the first hardware synchronization mechanism integrated in COOL, because it needs no bus access for synchronization. However, the `state`-signal requires $\lceil log(N) \rceil$ bits to encode N different states. Therefore, in some cases (e.g. if the ASIC has only a small number of pins available) it can be necessary to implement the hardware synchronization methods III or IV requiring only one or three pins for synchronization.

Table 5.2 summarizes these five synchronization schemes integrated in COOL.

No.	used for	started by	state updated by	#R/W	#IRQs	#pins
I.	SW	`irq`	system controller	3	1	0
II.	SW	`irq`	processor	0	1	2
III.	HW	`start`	system controller	3	0	1
IV.	HW	`start`	data path controller	0	0	3
V.	HW	`state`	system controller	0	0	$\lceil log(N) \rceil + 1$

Table 5.2. Synchronization schemes supported by COOL

5.4.2 *I/O Controller*

Data flow dominated systems supported by COOL are *transformational systems* which work on input data and produce output data. An audio system reading in digital samples and producing output samples is an example of this. The task of the *I/O controller* is to write input signals from the environment into memory and to transfer the computed results from memory to the environment. The principle functionality is depicted in figure 5.14.

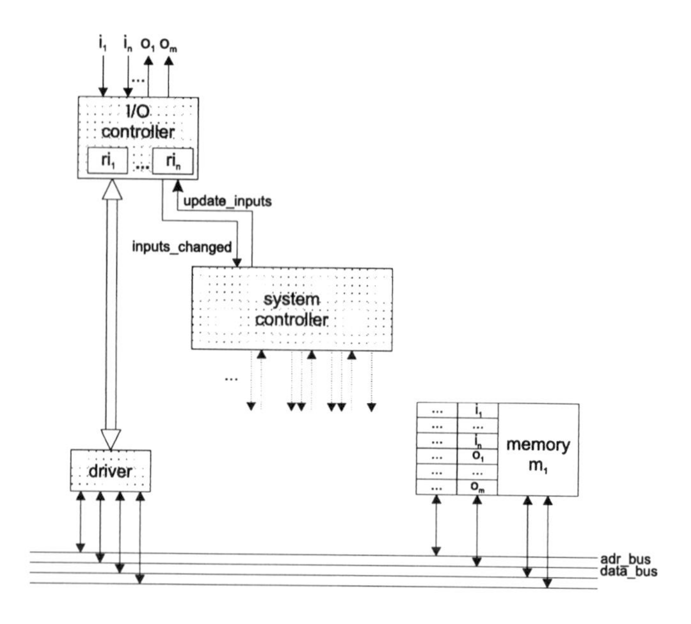

Figure 5.14. I/O controller

The I/O controller consists of two processes working concurrently: an *input* and an *output handler.*

1. *Input handler for writing inputs into memory and re-starting the system:* If an input signal i_n changes its value, the new value is always copied into an internal register ri_n for this input signal. In such a case, an internal flag **changed** is set, identifying that an input signal has changed. The new value is not written directly into memory, to prevent overwriting effects. These effects may occur if the system computes the result(s) for a given set of inputs, and if during this computation an input signal changes its value. For this reason, the signals which have changed are only written into memory when the system finishes the computation for the old inputs. A handshake mechanism implements this behavior: After the results for the old inputs have been computed, signal **update_inputs** is set to **'1'** by the system controller. If the internal **changed**-flag of the I/O controller is set, the inputs which have changed are written into memory. After this, the I/O controller sets signal **inputs_changed** to **'1'**. This indicates to the system controller that a new computation is necessary.

   The advantage of re-starting the system only when inputs have changed, implies a reduction in power consumption. However, in some cases when the computation of some output signals depends on the results of previous computations (e.g. accumulation), then the I/O controller should always set **inputs_changed** to **'1'** whether inputs have changed or not. Both possibilities have been integrated in COOL. The designer has to select the functioning of the I/O controller.

2. *Output handler for transferring the computed results to output signals*: For each output signal one or more memory cells are reserved. For this reason, the I/O controller contains a process implementing an output handler with the help of a *bus snooper*. Whenever the value on the address bus is equal to address `adr` which represents the memory cell storing output signal `o`, then the value of the data bus is copied to output signal `o`.

5.4.3 Bus Arbiter

The I/O controller, the system controller, all data path controllers ($\mathtt{FIR}_1$-, $\mathtt{FIR}_2$-controller in figure 5.15) and all processors may access the global memory using the bus. Therefore, a *bus arbiter* as depicted in figure 5.15 is necessary to prevent bus conflicts.

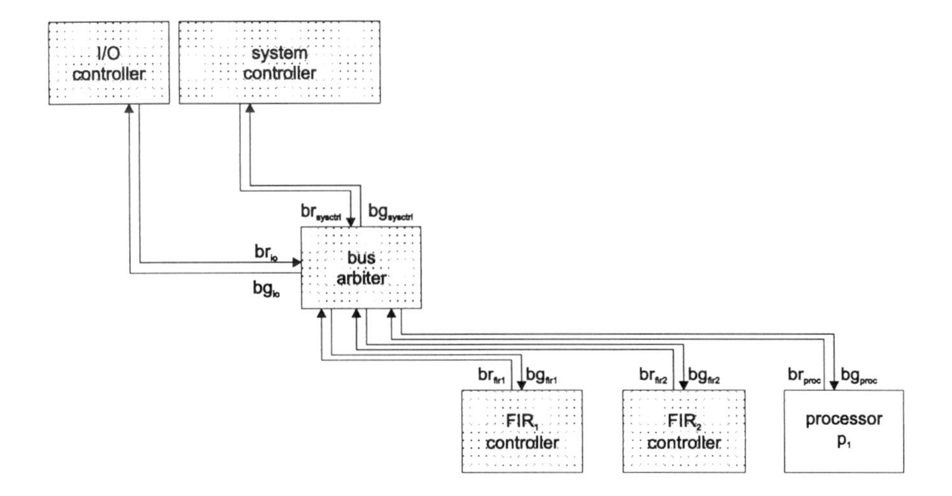

Figure 5.15. Bus arbiter

Each component may request the bus by setting its *bus request* signal to `'0'`, e.g. $\mathtt{br}_{IO}$=`'0'` for the I/O controller. The bus arbiter allows the I/O controller to use the bus which is indicated by setting the *bus grant* signal to $\mathtt{bg}_{IO}$=`'0'` if no other component uses the bus at this time. The bus arbiter implements a *fixed-priority arbitration scheme*. If more than one component requests the bus simultaneously, the bus arbiter assigns the bus to the component with the highest priority. The system controller has the highest priority to update the states of the processing units as soon as possible. Bus transfers are only interrupted if the system is initialized. Otherwise, a bus transfer is executed and after this the bus arbiter may grant the bus to other components.

5.4.4 Controller for Data Paths

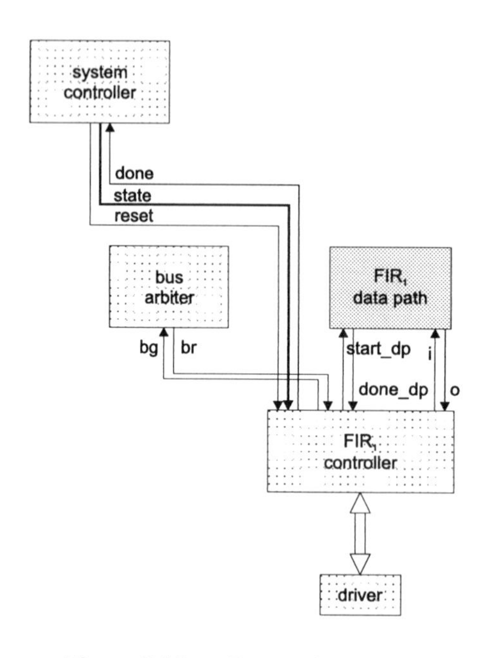

Figure 5.16. Data path controller

The data path controllers (see figure 5.16) represent the interface between the system controller steering the complete system and the data paths. For the sake of simplicity, the following description refers only to the fifth synchronization scheme presented in section 5.4.1. The data path controller is initialized by the system controller with the help of the `reset`-signal depicted in figure 5.16. After this initialization, the data path controller works as follows: First, the system controller starts the data path controller using the `state`-signal. Then, the data path controller requests the bus from the bus arbiter by setting `br='0'`. If the bus is available (`bg='0'`), all input signals required for the data path are read from memory. Then, the data path controller starts its data path using the `start_dp` signal. The data path computes the result and the `done_dp` signal is set afterwards. The data path controller requests the bus again (`br='0'`) and if the bus is available (`bg='0'`), the data path controller writes all results computed by the data path into memory. Finally, the `done`-signal is set to indicate to the system controller the end of computation.

5.4.5 Bus Drivers

As mentioned earlier, OSCAR does not support bi-directional signals and the Z-value of data type `std_logic` in `VHDL`. Therefore, it is impossible to connect designs synthesized with OSCAR directly to busses.

To solve this problem, *bus drivers* (see figure 5.17) are used to connect hardware components to busses. These bus drivers are synthesized with SYNOPSYS supporting bi-directional busses and the Z-value. If the bus is requested by the data path controller and granted by the bus arbiter (`br=bg='0'`), the bus driver sets all bus signals to the corresponding values of the signals coming in from the data path controller. Otherwise, all output signals are decoupled from the bus by setting them to `'Z'`.

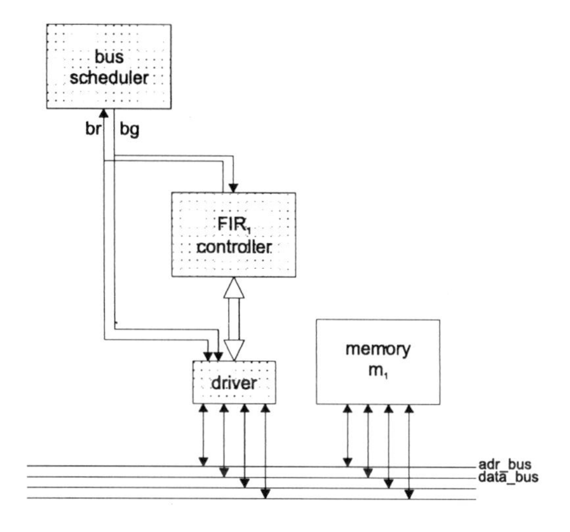

Figure 5.17. Bus driver

5.4.6 Simulation Models for Memory and Processors

Processors and memories are fixed elements of the target architecture in COOL. For them, therefore, only pre-defined simulatable **VHDL** descriptions are used from the **VHDL** package **cosys_pack.vhdl** (compare section 3.5). These **VHDL** entities are parameterizable and can be adapted to the chosen target architecture.

Global Memory:

Global memory is required to implement the communication between processing units via shared memory. Busses are used to connect the memory with the processing units.

Local Memory:

In some cases, the usage of *local memory* is of great advantage to reduce the transfer rates on busses. In COOL, local memory is always instantiated in the netlist if the data path internally operates on arrays, e.g. in an FFT[5]. In such a case, the local memory is directly connected to the data path (see figure 5.18) allowing the data path to access its data without conflicts with other processing units.

Processors:

Finally, processors are instantiated in the netlist. It has to be mentioned

[5]FFT: Fast Fourier Transformation

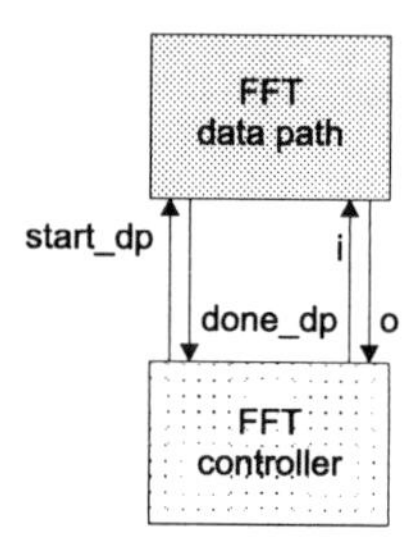

Figure 5.18. Local memory

that the processor model is a very simple VHDL description applicable to all processors having an interrupt and a memory interface. Simulating the software running on a processor is realized by simulating the processor model while exploiting the special features of the VANTAGE OPTIUM simulator. VANTAGE OPTIUM allows to call external C functions during simulation. Therefore, simulating the software is implemented by calling the generated software specification (in C) during hardware simulation of the simple VHDL description. As a negative consequence, the simulation can only validate functional correctness without being cycle-true.

5.5 State-Transition Graph

The previous section has described the general functioning of hardware/software systems implemented by COOL. This section shows how these hardware/software systems can be generated automatically. The basic idea for generating the controllers (system controller, data path controllers) is to divide the stages of computation for each node v of the partitioning graph into three *states* executed sequentially:

1. **WAIT**: In this state, node v waits to be executed until all input data and the resource (processor or hardware instance) is available. All input data can be accessed if all predecessor nodes w of v have computed their results. The resource R is available for v if all nodes scheduled on R before v have finished execution.
2. **EXECUTE**: During this state the computation for v is executed.
3. **DONE**: The computation for v has finished.

In figure 5.19, an example is given of these states described by the behavior of the processor.

Example 24:

The hardware/software partitioning algorithm has mapped nodes v_1 and v_3 to a hardware instance $h_{1,1,1}$ (data path and its controller) while v_2 has been mapped to processor p_1. The resulting hardware/software implementation works as follows: At the

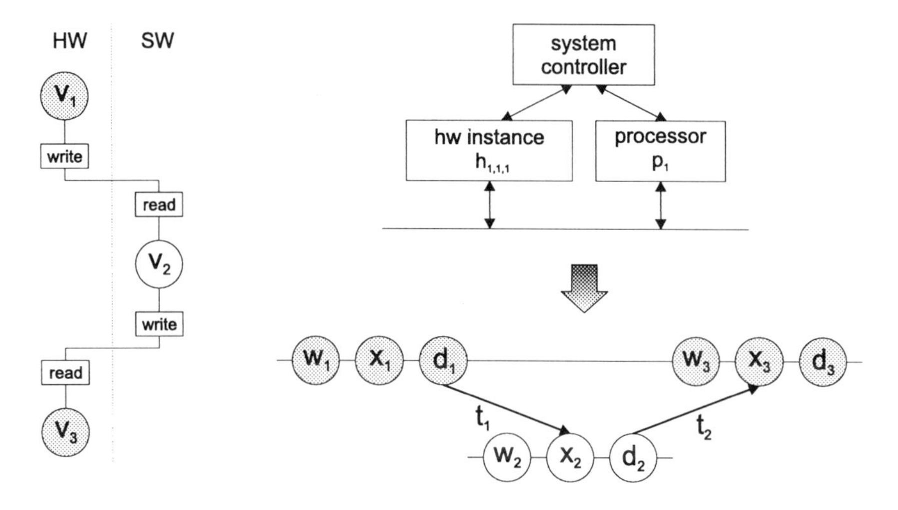

Figure 5.19. States and transitions

beginning, both processing units are waiting to start execution. Thus, hardware instance $h_{1,1,1}$ stays in state w_1 and processor p_1 stays in state w_2. Then, $h_{1,1,1}$ is started and its state changes from w_1 to x_1. When the result of v_1 is computed, hardware instance $h_{1,1,1}$ changes from state x_1 to d_1. Now, the system controller is able to activate processor p_1, because both predecessor states of x_2 (d_1 and w_2) are active. Thus, the processor leaves state w_2 and changes to state x_2 and in parallel the hardware instance changes from state d_1 to w_3 waiting to execute v_3. In state x_2, processor p_1 computes the results for v_2. If computation finishes, p_1 changes to state d_2. In this state, v_2 sends a message to the system controller using transition t_2. Finally, $h_{1,1,1}$ is started again by the system controller to execute v_3.

In the following, the generation of the state-transition graph for a given colored partitioning graph will be described.

5.5.1 *State-Transition Graph Generation*

All target architecture components may work concurrently. This feature is supported by the fact that the control of the system is not centralized, but distributed. Each processor contains a small control program in software and each data path has its own data path controller for preparing the computation. In addition, the system controller implements the interactions between different processing units. To compute cost-minimized implementations of these controllers, COOL uses a *state-transition graph* (STG) as an intermediate representation. The STG models the state of the system in a Petri net like manner. Therefore, the state of the complete system is defined by the set of marked states of the STG at a time. The STG is generated using the mapping and scheduling results from hardware/software partitioning. Afterwards, several optimizations are performed to minimize the number of states resulting in reduced controller costs. The state-transition graph is defined as follows:

Definition 14: *State-transition graph*

Let $G^p = (V^p, E^p, C^p, I^p)$ *be a partitioning graph.*
$\mathcal{T} = (\mathcal{V}, \mathcal{E})$ *be a target technology with* $\mathcal{V} = \mathcal{H} \cup \mathcal{P} \cup \mathcal{M}$ *(def. 2)*
$p_k \in \mathcal{P}$ *be a processor and* $h_k \in \mathcal{H}$ *a hardware component.*
$Map(G^P, \mathcal{T}) = (Map_\mathcal{V}, Map_\mathcal{E})$ *be a mapping of* G^P *onto* T *(def. 7).*

A **state-transition graph** *(STG) is a directed acyclic graph* $G^{st} = (V^{st}, E^{st})$*. Each node* $v \in V^{st}$ *represents either*

- *a global system state* $S \in \{R, X, D\}$ *(reset, execution, done), or*
- *a reset state* r_k *for a processor* p_k*, or*
- *a reset state* $r_{j,l,k}$ *for a hardware implementation* $h_{j,l,k} \in Map_\mathcal{V}$*, or*
- *an execution state* $s \in \{w_i, x_i, d_i\}$ *(wait, execution, done) for a node* $v_i^p \in G^p$*.*

Each edge $e = (v_1, v_2) \in E^{st}$ *indicates that state* v_1 *has to be active to activate state* v_2*. In general, a state* v_2 *can be activated if all predecessors* v_1 *of* v_2 *for edges* $e = (v_1, v_2) \in E^{st}$ *are active.*

Algorithm 5 computes a state-transition graph for a given partitioning graph.

Algorithm 5: **PGraph2STGraph**

```
(1)  algorithm PGraph2STGraph;
(2)       input G^p = (V,C,E,I)       : PartitioningGraph;
(3)       input T                     : TargetTechnology;
(4)       input Map(G^P,T)            : HWSWMapping;
(5)       output G^st = (V^st,E^st)   : StateTransitionGraph;
(6)  Let
(7)       V = V^I ∪ V^RI ∪ V^WO ∪ V^R ∪ V^W
(8)       Map(G^P,T) = (Map_V, Map_E)
(9)  {
(10)      variable p_k                                 : Processor;
(11)      variable h_j,l,k                             : HardwareInstance;
(12)      variable v_i^p, v_i1^p, v_i2^p               : node of PartitioningGraph;
(13)      variable L                                   : list of nodes;
(14)      variable e_i^p                               : edge of PartitioningGraph;
(15)      variable R, X, D, r_k, r_j,l,k, w_i, x_i, d_i : node of StateTransitionGraph;
(16)      variable x_i1, d_i1, w_i2, x_i2, d_i2        : node of StateTransitionGraph;
(17)
(18) // Create States
(19)      G^st.InsertStates(R,X,D);
(20)      forall processors p_k do G^st.InsertState(r_k);
(21)      forall hardware instances h_j,l,k ∈ Map_V do G^st.InsertState(r_j,l,k);
(22)      forall v_i^p ∈ V^I do G^st.InsertStates(w_i, x_i, d_i);
(23) // Create Transitions
(24)      forall v_i^p ∈ V^I do
(25)      {
(26)           G^st.Get_All_States_For_Node(v_i^p, w_i, x_i, d_i);
(27)           G^st.Insert_Transition(w_i, x_i);
```

(28) $G^{st}.Insert\_Transition(x_i, d_i);$
(29) }
(30) **forall** $e_i^p = (v_{i_1}^p, v_{i_2}^p) \in E$ **do**
(31) **if** $(v_{i_1}^p \in V^{WO} \cup V^W) \wedge (v_{i_2}^p \in V^{RI} \cup V^R)$ **then**
(32) **if** $Different\_Resources\_Used(v_{i_1}^p, v_{i_2}^p)$ **then**
(33) {
(34) $d_{i_1} = G^{st}.Get\_DONE\_State(v_{i_1}^p);$
(35) $x_{i_2} = G^{st}.Get\_EXEC\_State(v_{i_2}^p);$
(36) $G^{st}.Insert\_Transition(d_{i_1}, x_{i_2});$
(37) }
(38) $G^{st}.Get\_Global\_States(R, X, D);$
(39) **forall** processors p_k **do**
(40) {
(41) $r_k = G^{st}.Get\_RESET\_State(p_k);$
(42) $v_{i_1}^p = G^p.Get\_First\_Node\_Scheduled\_On(p_k);$
(43) $v_{i_2}^p = G^p.Get\_Last\_Node\_Scheduled\_On(p_k);$
(44) $w_{i_1} = G^{st}.Get\_WAIT\_State(v_{i_1}^p);$
(45) $x_{i_1} = G^{st}.Get\_EXEC\_State(v_{i_1}^p);$
(46) $d_{i_2} = G^{st}.Get\_DONE\_State(v_{i_2}^p);$
(47) $G^{st}.Insert\_Transition(R, r_k);$
(48) $G^{st}.Insert\_Transition(r_k, w_{i_1});$
(49) $G^{st}.Insert\_Transition(X, x_{i_1});$
(50) $G^{st}.Insert\_Transition(d_{i_2}, D);$
(51) $L = G^p.Get\_Nodes\_Implemented\_On(p_k);$
(52) $Sort\_Nodes\_By\_Schedule(L);$
(53) **forall** $v_{i_1}^p \in L$ **do**
(54) {
(55) **if** $v_{i_1}^p \neq L.head()$ **then**
(56) {
(57) $d_{i_1} = G^{st}.Get\_DONE\_State(v_{i_1}^p);$
(58) $w_{i_2} = G^{st}.Get\_WAIT\_State(v_{i_2}^p);$
(59) $G^{st}.Insert\_Transition(d_{i_1}, w_{i_2});$
(60) }
(61) $v_{i_2}^p = v_{i_1}^p;$
(62) }
(63) }
(64) **forall** hardware instances $h_{j,l,k}$ **do**
(65) {
(66) ...
(67) }
(68) }

The following example illustrates the algorithm. A state-transition graph for a partitioned equalizer is computed, as depicted in figure 5.20.

Example 25:

First, the algorithm inserts three global system states into the STG in line 19. These states are R (reset system), X (execute system) and D (system execution has been done). The computed hardware/software partitioning depicted in figure 5.20 contains two hardware instances FIR$_1$ and FIR$_2$ of an FIR-filter. FIR$_1$ executes nodes v_1 and v_2 whereas FIR$_2$ computes the results for nodes v_3 and v_4. The processor executes the multiplications $(v_5, \ldots, v_8)$ and the additions $(v_9, \ldots, v_{11})$. For each processor

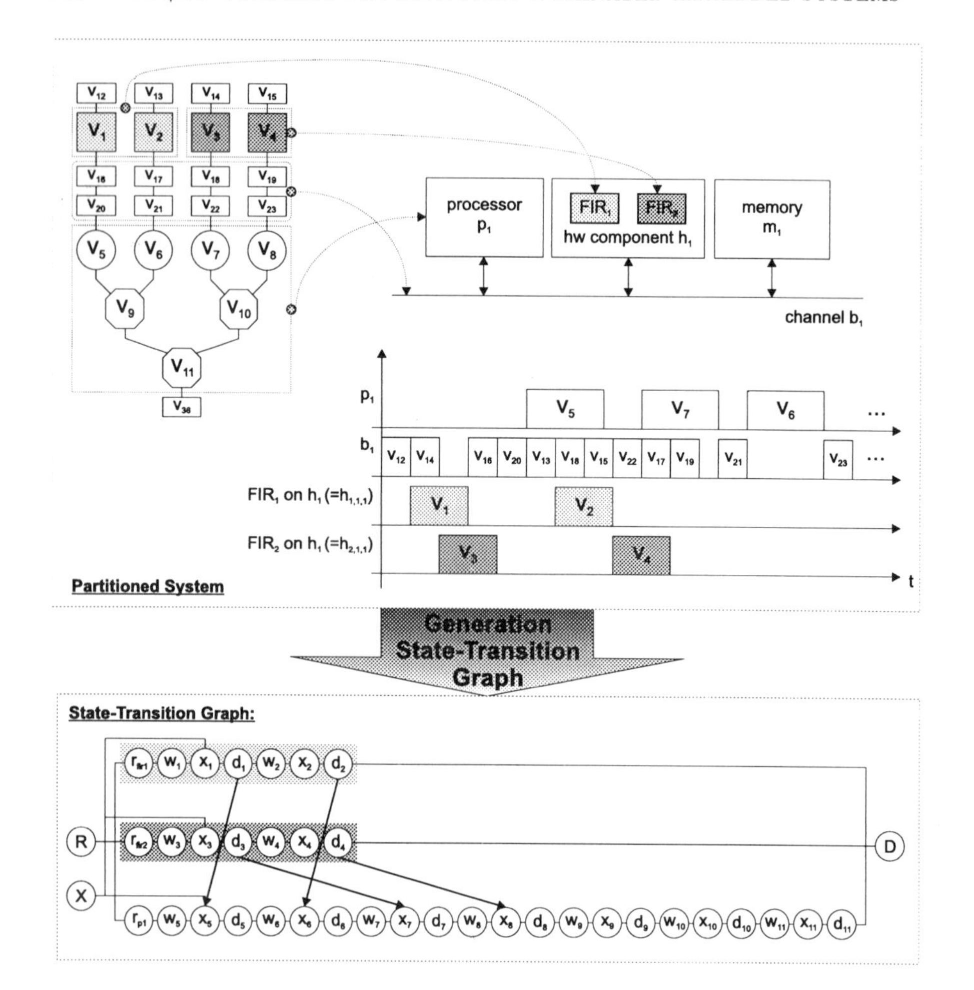

Figure 5.20. State-transition graph (STG) generation

and hardware instance of the target architecture (p_1, $\texttt{FIR}_1$, $\texttt{FIR}_2$) a RESET-state ($r_{p1}, r_{fir1}, r_{fir2}$ in figure 5.20) is added (see lines 20 and 21). For each node v_i of the partitioning graph three states are generated in line 22 corresponding to definition 14. Thus, 7 states are necessary to control hardware instance $\texttt{FIR}_1$, 7 states for $\texttt{FIR}_2$, and 22 states for processor p_1. However, this number of states can be minimized which will be described in section 5.5.2. The edges are generated in lines 24 to 67. First, the edges for realizing the execution order (wait, execute, done) of the three states are inserted in lines 24 to 29. Then, the edges required for synchronization between different processing units (with the help of the system controller) are inserted in line 36. These edges ($d_1 \rightarrow x_5, d_2 \rightarrow x_6, d_3 \rightarrow x_7, d_4 \rightarrow x_8$) are bold in figure 5.20. In lines 38 to 63, the edges connecting the states of each processor are inserted. The global system RESET-state R is connected to the RESET-state r_k in line 47. In line 48 the RESET-state r_k itself is connected to the WAIT-state w_{i_1} of node v_{i_1}, being the first node scheduled on processor p_k. In addition, the global EXECUTION-state X is connected to EXECUTION-state x_{i_1} of v_{i_1} (see line 49). The DONE-state d_{i_2} of the last scheduled node v_{i_2} is connected to the global system DONE-state D in line

50. In lines 51 to 62, the edges representing the local schedule on each processor are inserted.

With the help of figure 5.20, the functionality of these edges will be explained. Edge $R \rightarrow r_{p1}$ ensures that a global system reset initializes the processor. Edge $r_{p1} \rightarrow w_5$ guarantees that processor p_1 is waiting to execute v_5 until it is started by the system controller. Edge $X \rightarrow x_5$ determines that the processor starts executing v_5 when the complete hardware/software system is started. Finally, edge $d_{11} \rightarrow D$ indicates to the system controller that processor p_1 has finished execution.

For the sake of simplicity, the description of generating edges required for hardware instances has been omitted (see lines 64 to 67 in algorithm 5).

5.5.2 *State-Transition Graph Optimization*

After the STG has been computed, it is optimized to reduce the design costs for implementing the controllers based on the STG. These optimizations include the elimination of redundant transitions and states.

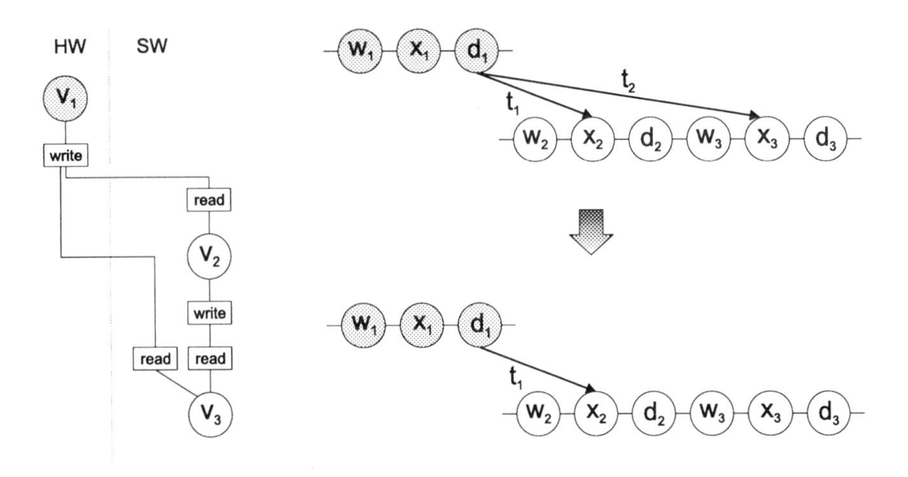

Figure 5.21. Transition elimination for the STG

5.5.2.1 Elimination of Transitions. In a first step, transitions included in the STG which are not necessary are eliminated. In figure 5.21, an example is given. Node v_1 is implemented on a processing unit different from the one used by v_2 and v_3. The computed result of v_1 is required by v_2 and v_3. In addition, v_3 requires the result computed by v_2. Transition t_1 represents the activation of v_2 and t_2 represents the activation of v_3 by the system controller. In such a case, only t_1 is necessary for synchronization, because v_3 has to wait for the result of v_2. If v_2 has finished execution, it is guaranteed that v_1 has finished before. Therefore, transition t_2 can be eliminated.

5.5.2.2 Elimination of States. Figure 5.22 illustrates the elimination of redundant states leading to a significant reduction of the design costs.

Example 26:

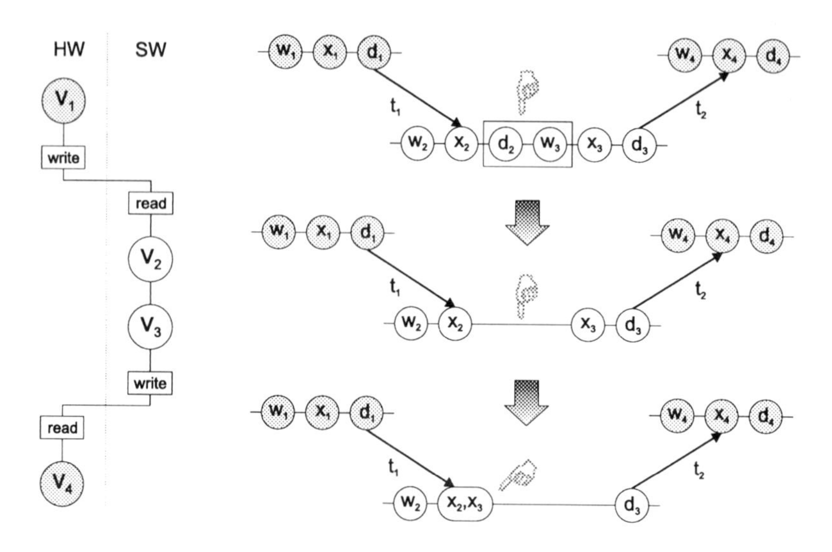

Figure 5.22. State elimination for the STG

Transition t_1 activates state x_2 if v_1 has finished execution. Then, v_2 and v_3 are executed sequentially on the same processing unit. The results computed by v_2 are only used by v_3. Therefore, the result has not to be stored in external memory. In addition, v_3 requires only the results computed by v_2. For this reason, v_3 may start directly after v_2 finishes without additional synchronization by the system controller. Summarizing, states d_2 and w_3 used for implementing the synchronization with the system controller are eliminated. In this case, states x_2 and x_3 can be merged to one state.

The following general observation can be made with respect to this example:

- A WAIT-state w_i is necessary if a node v_i requires data computed on another processing unit. In such a case, the system controller has to synchronize the activation of x_i, using state w_i, until all input data for node v_i is available.
- A DONE-state d_i is necessary if the results computed for node v_i are consumed on another processing unit. In such a case, state d_i indicates to the system controller that v_i has computed its results.
- An EXECUTION-state x_i is necessary if one of its predecessor states is required.

Based on these observations, WAIT- and DONE-states can be eliminated for the cases described before. In the following example, the presented optimization techniques have been applied to the STG computed for the equalizer.

Example 27:

The elimination of transitions has no effect for the state-transition graph depicted in figure 5.23. However, the elimination of states could be applied to the equalizer. Seven

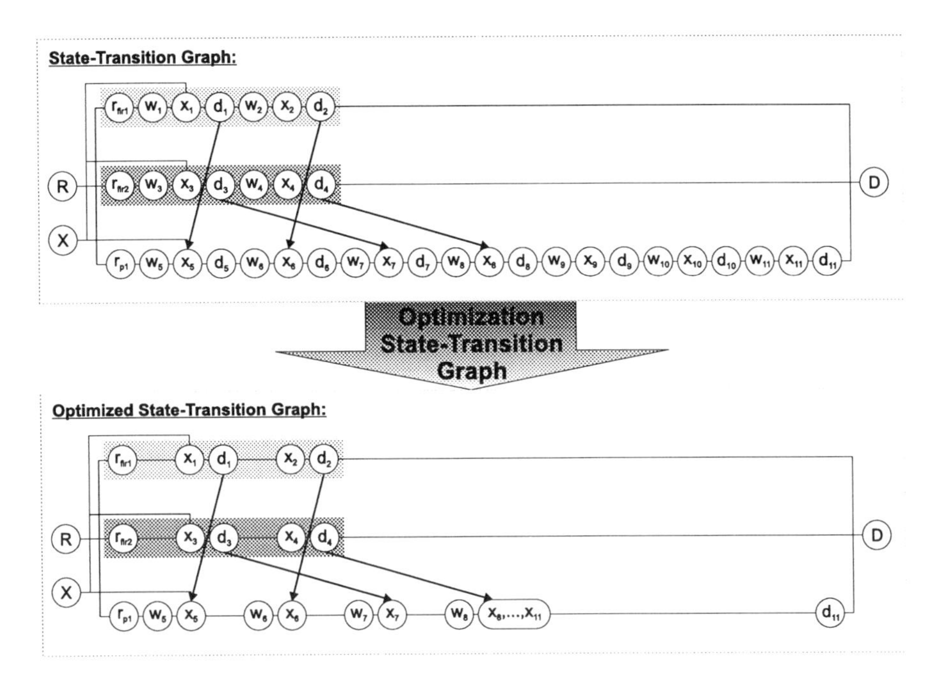

Figure 5.23. Example of STG optimization

WAIT-states $w_1, \ldots, w_4, w_9, \ldots, w_{11}$ and six DONE-states $d_5, \ldots, d_{10}$ are removed. In addition, the execution states $x_8, \ldots, x_{11}$ are merged to one state, because they can be executed in sequence without synchronization by the system controller. This reduces the number of states from 39 to 23.

5.5.3 *Memory Allocation*

The next step for generating the controllers is to allocate memory cells of the shared memory starting from a user-defined *base* address to enable *memory mapped I/O*.

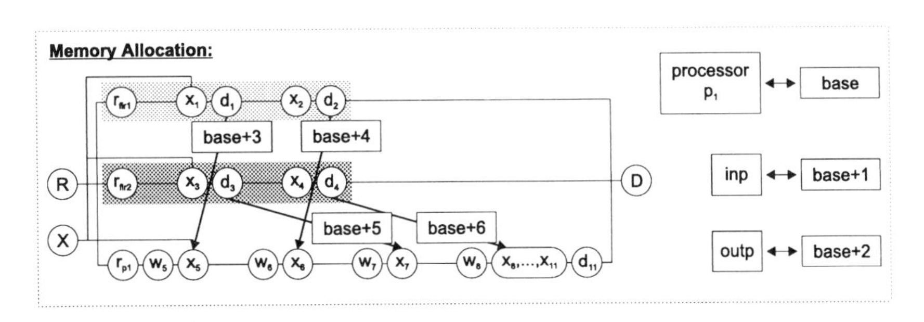

Figure 5.24. Memory allocation

Three different tasks are performed during memory allocation by COOL as illustrated in figure 5.24. Let the data width of the memory be defined as $N_{datawidth}$.

1. As described in section 5.4.1, synchronization between the system controller and processing units may be implemented using *shared memory* if synchronization schemes I or III have been selected. In such a case, the state of the processing unit is stored in memory (see address `base` in figure 5.24). Thus, for each processing unit executing N_{states} states n_1 memory cells have to be allocated for writing and reading the state:

$$n_1 = \left\lceil \frac{log_2(N_{states})}{N_{datawidth}} \right\rceil \quad (5.1)$$

   If synchronization schemes I or III have not been selected, then this step can be ignored.

2. The I/O controller writes incoming signals into memory (address `base+1`) and reads outgoing signals (address `base+2`) from it. Therefore, memory cells have to be allocated for these signals. To store a signal consisting of N_{bits} bits, n_2 memory cells are required. The number n_2 is computed by

$$n_2 = \left\lceil \frac{N_{bits}}{N_{datawidth}} \right\rceil \quad (5.2)$$

3. For all processing units exchanging data memory cells have to be allocated (see memory cells `base+3`,...,`base+6` in figure 5.24). Thus, for all transitions connecting the DONE-state of node v_1 with the EXECUTION-state of v_2 memory cells have to be allocated for the data produced by v_1 and consumed by v_2. To transport the data using a communication node (the data type should consists of of N_{bits} bits), n_2 memory cells are required too (see eq. 5.2).

5.6 Refinement and Controller Generation

After the STG has been optimized and the memory cells have been allocated, the original specifications are refined and the additional controller descriptions are generated. First, some data-related refinement steps will be described in section 5.6.1 which are performed during both hardware and software refinement. Then, software refinement will be presented in section 5.6.2 and hardware refinement in section 5.6.3. Finally, the generation of the controller descriptions will be introduced in section 5.6.4.

5.6.1 Data-Related Refinements for Hardware and Software

Refinement steps translate the implementation-independent system specification by replacing certain parts through target-specific information. This section describes transformations for both hardware and software. As described in section 5.1.2.2, data-related refinement includes the following tasks:

- data type conversion,
- operation replacement,
- variable folding and
- memory address translation.

5.6.1.1 Data Type Conversion. The types included in the abstract VHDL specification have to be replaced by suitable types of the implementation language. The following types (see table 5.3) defined in COOL are replaced by corresponding type declarations in VHDL for the hardware parts and C for the software parts.

Type in COOL	Hardware (synthesizable VHDL)	Software (C)
bit	std_logic	char
unsigned_int_16	std_logic(15 downto 0)	unsigned
unsigned_fix_16_8	std_logic(15 downto 0)	unsigned
array_signed_fix_32_16_8	std_logic(15 downto 0)[6]	signed[32]

Table 5.3: Data type conversion for system specification

Hardware synthesis tools are often not able to handle arrays declared in a specification. For this reason, arrays are considered by COOL with the help of local memories. The type of an array variable is transformed corresponding to the base type of the array type. More details will be given in section 5.6.3.

5.6.1.2 Operation Replacement. In table 5.3, it is shown that types unsigned_int_16 and unsigned_fix_16_8 are transformed into the same type during hardware and software refinement. A special problem occurs in this case, because some operations defined for these types need some additional information which is lost during type conversion. Therefore, operations of different types have to be implemented differently in some cases. Otherwise, incorrect results would be computed. The typical example is fixed-point multiplication. To compute the correct result, the fraction of the fixed-point number has to be considered when multiplying. During the conversion of the fixed-point type this information is lost. An example of this fixed-point multiplication is shown in figure 5.25.

Example 28:

In figure 5.25, two multiplications of the same bit vectors are shown. The first multiplication multiplies two numbers of type unsigned_int_10. The result of this unsigned

[6] Arrays are not supported by OSCAR. Refinement uses local memory.

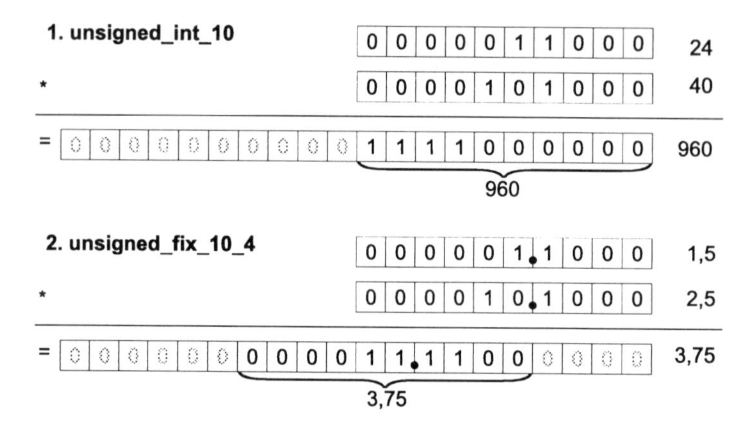

Figure 5.25. Operation replacement

multiplication is $24 * 40 = 960$. In this case, the lower 10 bits represent the result. For the fixed-point multiplication of type `unsigned_fix_10_4`, having a fraction of 4 bits, the result is $1.5 * 2.5 = 3.75$. It is computed with a standard unsigned multiplication followed by right-shifting the result by 4 bits.

Therefore, for each operation computing results for a special type an appropriate function has to be implemented if the standard operator of the target language is not sufficient. In table 5.4 some typical examples of operation replacement performed by COOL are given.

Example 29:

No.	Specification in COOL	Hardware (synth. 'OSCAR'-VHDL)
1	`a1 := CopyBits(a2,3,0);`	`a1 := a2(3 downto 0);`
2	`b1(1) := '1';`	`b1 := b1 or "0000000000000010";`
3	`c1 := c2 and c3;`	`c1 := c2 and c3;`
4	`d1 := Fix_Mul_16_8(d2, d3);`	`d1 := ufix_mul_16_8(d2, d3);`

No.	Specification in COOL	Software (C)
1	`a1 := CopyBits(a2,3,0);`	`a1 = a2 & 15;`
2	`b1(1) := '1';`	`b1 = b1 | 2;`
3	`c1 := c2 and c3;`	`c1 = c2 & c3;`
4	`d1 := Fix_Mul_16_8(d2, d3);`	`d1 = ((d2 * d3) >> 8) & 65535;`

Table 5.4. Examples of operation replacement

The first example illustrates the replacement of the '`CopyBits`'-operator (specially defined function in `VHDL` package `cosys_pack.vhdl` for copying bits). On the software side, copying bits is implemented by using a masking value. This is done for setting a single bit of a variable during hardware and software refinement too. The third example is a simple translation of the '`and`'-operator in `VHDL` into the '`&`'-operator in `C`. In other cases, the refinement can be more difficult as illustrated by the last example. Here, a fixed-point multiplication uses a specially defined hardware component stored in the component library of OSCAR. On the software side, the fixed-point multiplication is refined by standard multiplication combined with right-shifting and masking.

5.6.1.3 Variable Folding. The implementation-independent system specification contains signal declarations in the `port map` of the `entity` declaration. During hardware/software partitioning, some of these signals are mapped to communication channels. In such a case, the data is buffered in a shared memory which is connected by busses with the processing units. Therefore, the signals have to be folded as described in section 5.1.2.2. In table 5.5, all cases are depicted for reading/writing a folded signal v (with bit width n) using data bus db (with bus width N) from shared memory.

Case	i	**Read data v from data bus db**
1. $n < N$	0	$v = db.(n-1,\ldots,0)$
2. $n = N$	0	$v = db$
3. $n > N$	0	$v(N-1,\ldots,0) = db$
	$\ldots$	$\ldots = db$
	i	$v((i+1)*N-1,\ldots,i*N) = db$
	$\ldots$	$\ldots = db$
3a. $N = m*n$	$m-1$	$v(n-1,\ldots,(m-1)*N) = db$
3b. $N \neq m*n$	$m-1$	$v(n-1,\ldots,(m-1)*N) = db.(N-n-1,\ldots,0)$

Case	i	**Write data v onto bus db**
1. $n < N$	0	$db = \underbrace{0\ldots0}_{N-n}\&v$
2. $n = N$	0	$db = v$
3. $n > N$	0	$db = v(N-1,\ldots,0)$
	$\ldots$	$db = \ldots$
	i	$db = v((i+1)*N-1,\ldots,i*N)$
	$\ldots$	$db = \ldots$
3a. $N = m*n$	$m-1$	$db = v(m*N-1,\ldots,(m-1)*N)$
3b. $N \neq m*n$	$m-1$	$db = \underbrace{0\ldots0}_{N-n}\&v(n-1,\ldots,(m-1)*N)$

Table 5.5. Folding of signals into memory accesses

If n is smaller than N (case 1), then only the lower n bits have to be read. A write access prepends some '0's for the upper bits which are not required. Case 2 is very simple, because the bit widths of the data bus and the transported data are equal. If $n > N$ (case 3), then the data is iteratively read and written. Case 3b differs from case 3a in the way that the last read/written word uses only a part of the data bus, similar to case 1. It should be mentioned that in case 3 the data is read and written using *little endian* assignment (see section 5.1.2.2), but COOL is also able to store the data in *big endian* order.

5.6.2 Software Refinement

The task of *software refinement* in COOL (depicted in figure 5.26) is to refine the implementation-independent system specification corresponding to the

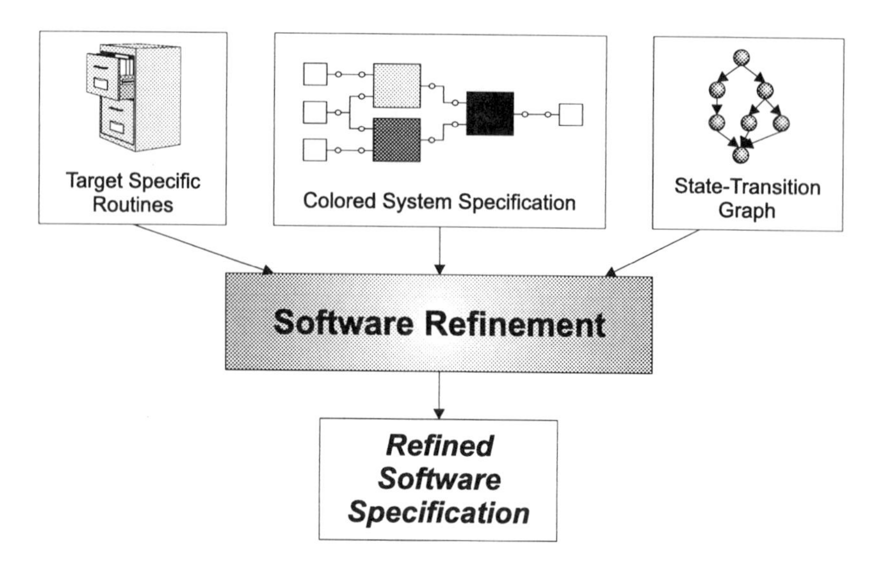

Figure 5.26. Software refinement in COOL

results of hardware/software partitioning into a compilable software specification. Therefore, COOL generates a new C file for each processor including all refined functions mapped to this processor. In addition, a small control program executes the functions corresponding to the computed schedule which is encoded by the state-transition graph. To be able to generate C files for different processors, COOL uses a combination of automatic refinement and *library management*. For each processor of the target architecture library the following target-specific functions have to be defined:

- a `read`- and a `write`-routine for reading/writing one word using the memory interface,
- a routine (`InitSettings`) for initializing the processor (e.g. for setting the number of wait-states to access external RAM),
- special procedures to access unused pins required for the software synchronization scheme II which has been presented in section 5.4.1.

With the help of these target-specific routines, COOL is able to generate C specifications for different processors. For each processor the C file is adapted by exchanging the included target-specific header file. In figure 5.27, an example is given of target-specific functions and a refined software specification using these functions.

The following list summarizes all parts of a refined C specification.

- Special `READ`- and `WRITE`-operations are included to access the external memory using the target-specific operations `read_word` and `write_word`. The

```
/*========================================
 Target-specific routines for Motorola DSP56001
========================================*/

void read_word(int *val, int adr)
{ __asm("move %1,r1\n\
        move x:(r1),%e0":"=D"(*val):"R"(adr):"r1");
}

void write_word(int *val, int adr)
{ __asm("move %1,r1\n\
        move %0,x:(r1)"::"S"(*val),"R"(adr):"r1");
}

void InitSettings()
{ __asm("movep #$0c06,X:$ffff\n\
        movep #$bb02,X:$fffe");
}

void SetUnusedPin(int n, int value) { ... }

int GetUnusedPin(int n) { ... }
```

```
/*========================================
 Refined software specification
========================================*/
#include "motorola56001.h"  /* target-specific routines */

typedef int unsigned_int_8;

void read_unsigned_int_8(unsigned_int_8 *val, int adr)
{ read_word(val, adr);
  *val = *val & 255;
}

void fct1(unsigned_int_8 i, unsigned_int_8 *o) { ... }
void fct2(...) { ... }

void IRQ_ServiceRoutine()
{ unsigned_int_8    state, i1, o1;

    read_unsigned_int_8(state, 16384);
    switch state of
    { case 1 : read_unsigned_int_8(i1, 16385);
               fct1(i1,&o1)
               write_unsigned_int_8(o1, 16386);
               state := 2;
               break;
      case ...
    }
    write_unsigned_int_8(state, 16384);
}

int main()
{ InitSettings();
  while(1);
}
```

Figure 5.27. Target-specific operations and refined software

target-specific R/W-operations for the MOTOROLA DSP56001 themselves use special assembler directives.

- Functions computing fixed-point arithmetic are integrated.
- For each node mapped to the processor a function (e.g. `fct1`) is included specifying the behavior of the node.
- An interrupt service routine is generated which calls a function corresponding to the `state` of the processor read from external memory (synchronization scheme I described in section 5.4.1). The functionality of the interrupt service routine can be compared with the functionality of a data path controller which activates its data path according to the computed schedule. When executing `state 1` (see figure 5.27), first input `i1` is read, function `fct1` is executed for `i1` and after this the computed result `o1` is written into external memory. Finally, the state of the processor is updated and also written into external memory.
- The main program executes the target-specific initialization procedure for the MOTOROLA DSP56001, defining the interrupt priorities and the number of wait-states to access external memory.

5.6.3 Hardware Refinement

The *hardware refinement* step of a data path transforms all data types into `std_logic` types. Special operations, e.g. fixed-point multiplications, are replaced by special functions defined in the functions library of OSCAR. This replacement ensures that a suitable component is selected during high-level synthesis.

```
entity fct1 is
  port( i  : in unsigned_int_8;
        o  : out unsigned_int_8);
end fct1;

architecture body of fct1 is
begin
  P:  process(i)
      begin
        <statement 1,...,n>
      end;
end fct1;
```

⇨

```
entity fct1 is
  port( clk        : in std_logic;
        start_dp   : in std_logic;
        done_dp    : out std_logic;
        i          : in std_logic_vector(7 downto 0);
        o          : out std_logic_vector(7 downto 0));
end fct1;

architecture body of fct1 is
begin
  P:  process(i)
      begin
        wait until clk'event and clk='1';
        if start_dp = '1' then
          <statement 1,...,n>
          done_dp <= '1';
        else
          done_dp <= '0';
        end;
      end;
end fct1;
```

Figure 5.28. Hardware refinement

In general, three new signals (`clk, start_dp, done_dp`) are added. Signals `start_dp` and `done_dp` are required to implement a handshake mechanism between data path controller and data path as illustrated in figure 5.28. The data path controller starts the data path by setting `start_dp` to `'1'`. Then, the data path computes the results and after this it sets the `done_dp`-signal to `'1'`.

Array declarations of the system specification have to be refined, because OSCAR is not able to synthesize arrays. Local memories are instantiated to handle these array declarations. For this reason, additional signals (`lm_adr, lm_data, lm_rw, lm_ds`) are required to connect a local memory to the data path. An access to an array within the specification is transformed into a handshake protocol, such as depicted in figure 5.29, accessing the local memory.

The base addresses of the specified arrays in the local memory have been computed during memory allocation.

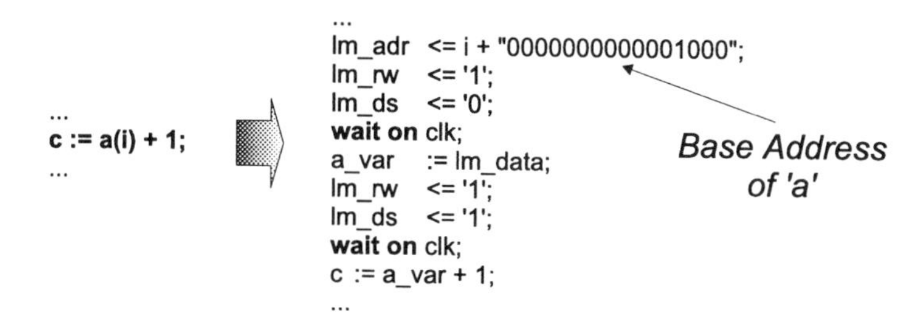

Figure 5.29. Transformation of arrays into local memory accesses

5.6.4 Controller Generation

For the *system controller* and the *data path controllers* finite-state machine (FSM) descriptions are generated in `VHDL`. The generation of the FSMs is based on the states and transitions encoded in the state-transition graph. The addresses allocated during the memory allocation step are now integrated in the refined specifications. The following example gives a short overview of the generated controller specifications for the 4-band equalizer.

Example 30:

In figure 5.30, some parts of the generated `VHDL` descriptions for the system controller and the data path controller for the first FIR-filter instance are shown.

First, the system controller checks an incoming `reset`-signal (state R). If `reset='0'`(!), the system is reset by initializing all processing units. This is done by *message passing synchronization* (`state_asic_1<="000", state_asic_2<="000"`) for the hardware components. In this example, processors are reset by *shared memory synchronization*. If `reset` changes to `'1'`, the first state "0000" of the finite-state machine is used to write the RESET-state of processor p_1 into memory at address `base`. Then, in state "0001" the interrupt signal of the processor is set to `irq_proc_1<='0'`. Thus, the processor will be reset. The system controller itself changes into state "0010" and blocks until the system is started by `start='1'`. In this case, the inputs are updated first by sending a message (`update_inputs<='1'`) to the I/O controller. If the inputs have not changed, the system controller stays in state "0010" to reduce power consumption (see section 5.4.2). However, if the inputs have changed, the state changes into "0011". In this state, both hardware instances of an FIR-filter are started by executing transitions $r_{fir1} \to x_1$ and $r_{fir2} \to x_3$. If the first FIR-filter has computed its results, checked in controller state "0110", then transition $x_1 \to d_1$ is executed. Afterwards, the processor can be started, etc.

The data path controller for the FIR-filter also checks first an incoming `reset` in form of `state="000"`. The data path controller blocks (staying in state "000") until the system controller sends an execution state (`state`$\neq$`"000"`) by message passing. In this case, the addresses for the in- and outputs are set depending on `state`. These addresses have been determined during memory allocation. After this, the inputs are read and the data path is started afterwards, etc.

This section only gives a short overview of the generated FSMs in form of `VHDL` descriptions. The interested reader is referred to [PG293b], where a complete listing of the generated netlist and all components described in `VHDL` and `C` for a partitioned fuzzy system is given.

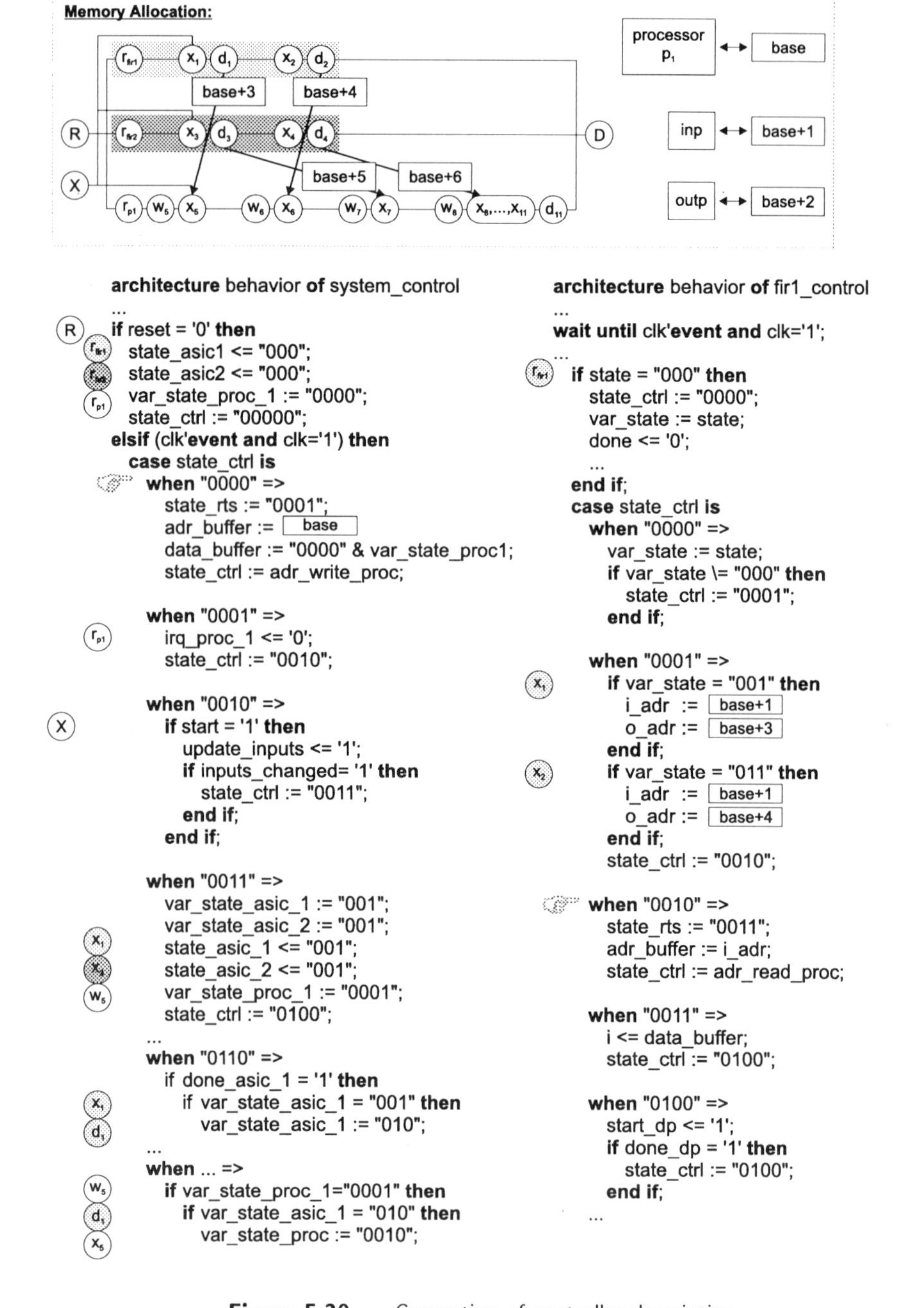

Figure 5.30. Generation of controller description

For the sake of simplicity, the `VHDL` code depicted in figure 5.30 does not contain memory accesses. The states for `WRITE`- and `READ`-accesses, implemented by a simple *handshake protocol*, are depicted in figure 5.31.

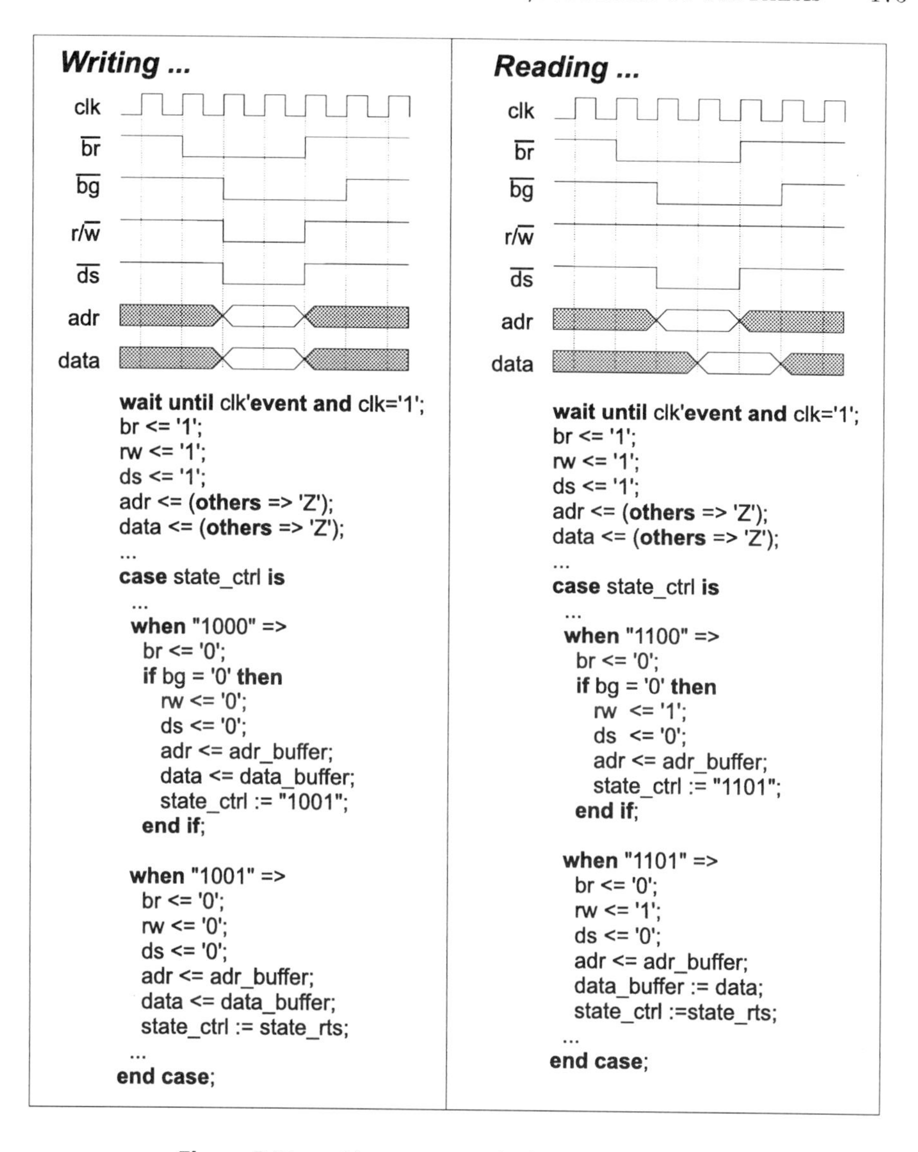

Figure 5.31. Memory access by hardware components

WRITE: The `WRITE`-operation starts with setting the *bus request* to `br='0'`. If the bus is available, the bus arbiter acknowledges by setting the *bus grant* signal to `bg='0'`. Then, the *data strobe* signal is set to '0' (`ds='0'`), indicating an access to memory. In parallel, the *read/write*-signal is set to '0' (`rw='0'`) defining a `WRITE`-access. Concurrently, *address bus* and *data bus* are set for two cycles. Finally, the bus is released by setting `br='1'`.

READ: The READ-operation works similarly. First, the bus is requested by br='0'. If the bus arbiter grants the bus (bg='0'), then the data strobe signal is set to '0' (ds='0'). The read/write-signal is set to '1' (rw='1') indicating a READ-operation and the *address bus* is set. In the depicted case, after one cycle the data can be read from the *data bus*. Finally, the bus is released by setting br='1'.

Note: In some cases additional *wait-states* are required to implement the memory accesses, depending on the memory chips.

All generated controller descriptions use these memory access states. To minimize the number of states, the sequences for WRITE- and READ-accesses can be specified by sub-routine calls. An example of calling the WRITE-subroutine is given in state "0000" of the system controller description, an example of a READ-access in the description of the data path controller depicted in figure 5.30. This READ-access works as follows: in state "0010" of the data path controller description, variable state_rts is set to "0011" representing the state executed after finishing the read-subroutine. Then, state_ctrl is set to adr_read_proc defining the controller state in which the read sub-routine starts. After the READ-subroutine has been executed, the FSM jumps into state state_rts in which the data input buffer is copied to signal i.

5.7 Results IV: Application Study of a Fuzzy Controller

COOL was used during a student project [PG293a, PG293b] to implement a fuzzy controller on a *heterogeneous target architecture*. The task of the fuzzy controller is to control the traffic lights at a certain crossing of a main street and a side street. This crossing contains three different lanes as depicted in figure 5.32.

The idea is to control the traffic lights depending on the number and the waiting time of cars in each lane. The goal of the fuzzy controller is to minimize the average latency of the cars. The principle functionality of the fuzzy controller is depicted in figure 5.33.

For each lane the fuzzy controller uses the same modules to compute an *urgency metric* to get the green-signal. To reuse the same modules for each lane allows to apply hardware sharing to minimize the total costs during the design process. The modules for each lane have the following inputs: the waiting time, the number of cars and a boolean value representing whether a pedestrian is waiting or not. Like most fuzzy controllers, these inputs (representing *linguistic variables*) are *fuzzified* in a first step computing *linguistic terms* for the inputs. Then, the *inference* applies some rules stored in a *rule base* to the linguistic terms. After this, the resulting values are *defuzzified* into discrete output values, here the urgency metric presented above. Finally, the lane with the highest value of this urgency metric gets the green-signal.

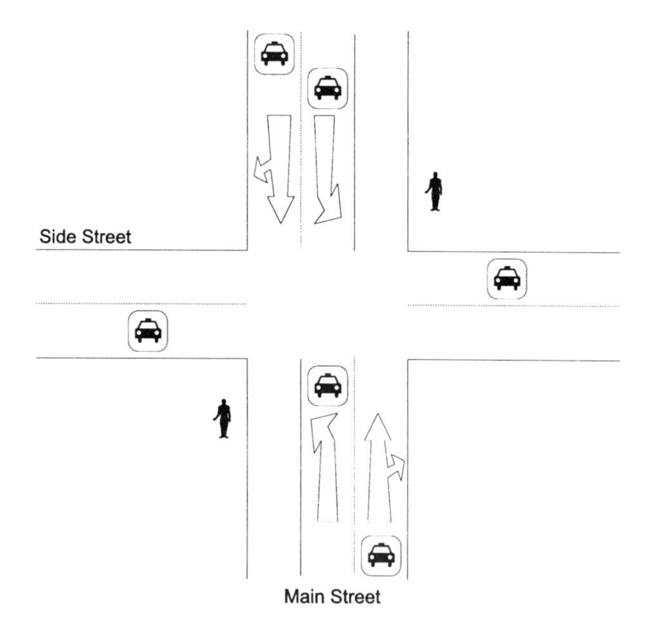

Figure 5.32. Crossing situation

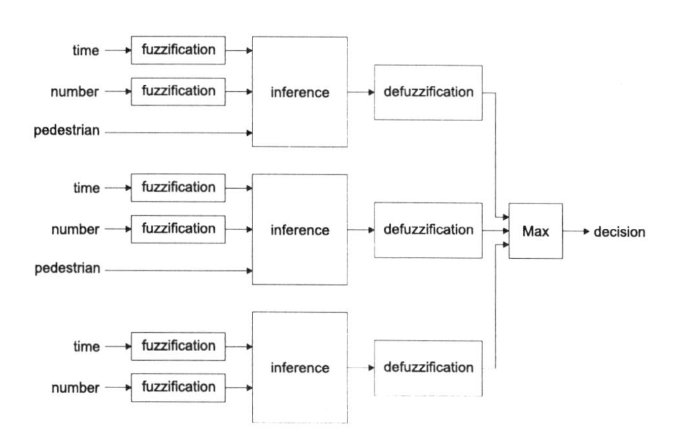

Figure 5.33. Block diagram of the fuzzy controller

First, the fuzzy controller was specified with COOL. The way of specifying systems in COOL is very well suited for this example, because the modules of the fuzzy controller can be specified by different system components which are instantiated for each lane afterwards. The resulting system specification consisting of 31 instances of 7 system components is depicted in figure 5.34.

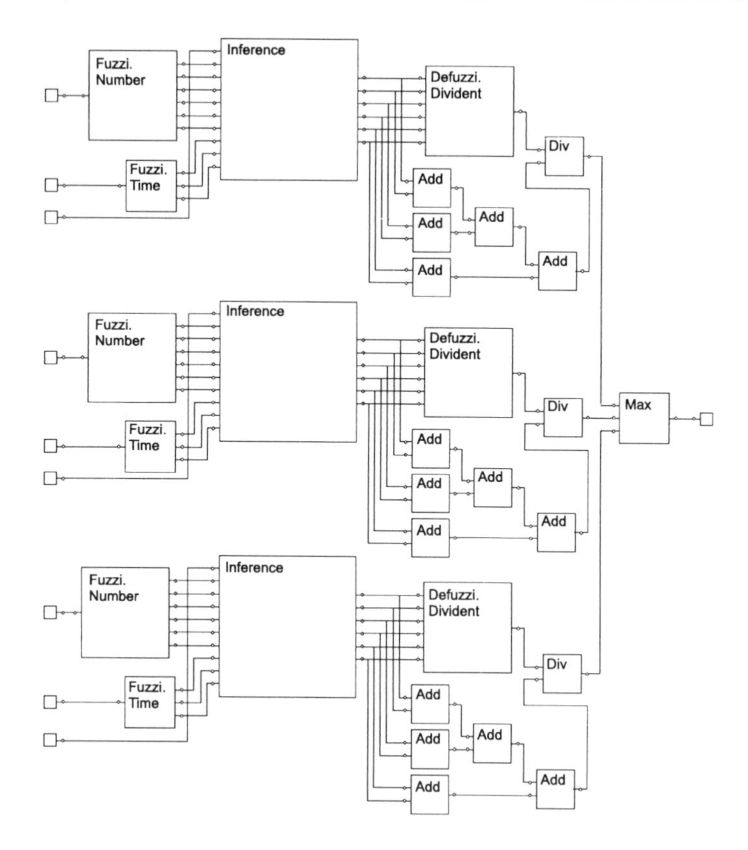

Figure 5.34. COOL specification of the fuzzy controller

In total, the behavioral VHDL specifications consist of 900 loc[7]. The goal was to implement this fuzzy controller specification on a target architecture as depicted in figure 5.35 containing

- a MOTOROLA DSP56001 [Moto94],
- an ASIC,
- a 64kB memory and
- a communication channel (connecting the DSP, the ASIC and the memory) consisting of an 8-bit data bus and a 16-bit address bus.

To validate the correctness of the resulting hardware/software system, the students developed a prototyping environment [PG293a], using

- a PC plug-in card from DAVIS [Davi93] containing the MOTOROLA DSP56001 and a sandwich card for the DAVIS card to connect the DSP to the busses,

[7]loc: lines of code

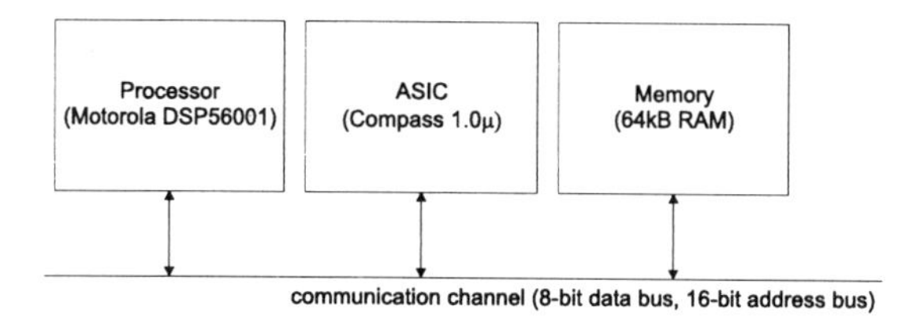

Figure 5.35. Heterogeneous target architecture for the fuzzy controller

- an FPGA-board with two XILINX FPGAs 4005 [Xili94] (with 196 CLBs[8] and 61 available pins each) to implement the ASIC and
- a memory board containing 64kB static RAM, additional bus drivers and RAM logic.

All these cards and boards were connected using an 8-bit data bus and a 16-bit address bus as depicted in figure 5.36. In figure 5.37, the prototyping board is depicted.

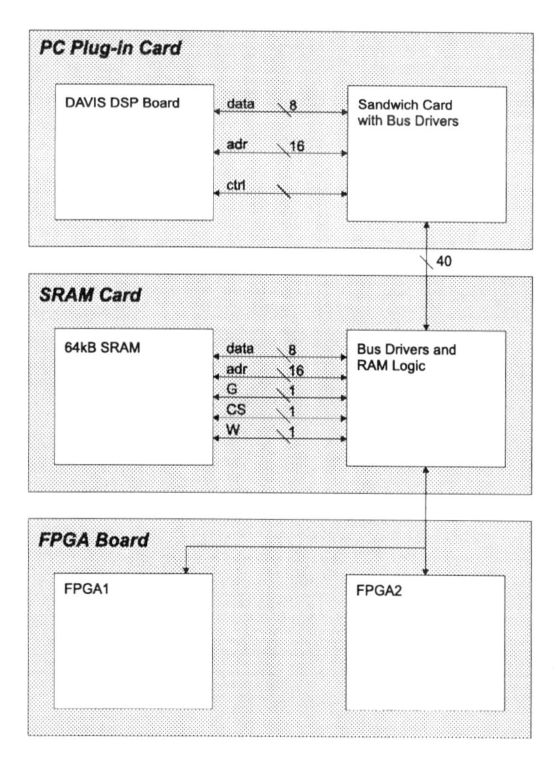

Figure 5.36. Prototyping environment to implement the fuzzy controller

[8]CLB: Configurable Logic Block

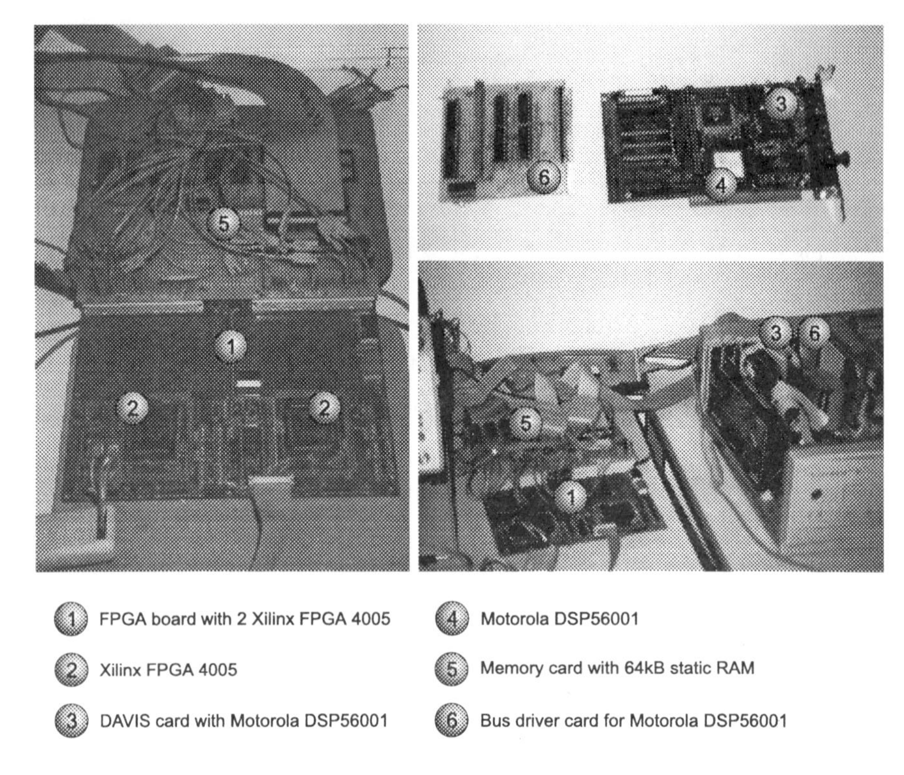

Figure 5.37. Prototyping board

The main disadvantage of this prototyping board was the limited amount of hardware resources with 2×196 CLBs and 2×61 pins. Most of the specified components of the fuzzy controller could not be implemented on the FPGAs due to the lack of CLBs. To implement the fuzzy controller on the prototyping environment, the hardware/software partitioning was driven by designer defined mapping constraints resulting in the following hardware/software partitioning, depicted in figure 5.38.

The computed partitioning consisted of 15 nodes implemented in hardware and 16 nodes implemented in software. Afterwards, all co-synthesis steps (generation and optimization of the STG, memory allocation, specification refinement, controller and netlist generation) were executed resulting in the netlist depicted in figure 5.39.

Then, the refined `C` specification for the MOTOROLA DSP (961 loc) was compiled into assembler code (5305 loc) and into a binary file afterwards. The hardware parts implemented on the ASICs were partitioned into two separated `VHDL` descriptions (one for each FPGA) manually, because otherwise pin constraints were violated. The system controller and the I/O controller were implemented on the first FPGA, the bus arbiter, the hardware data paths and

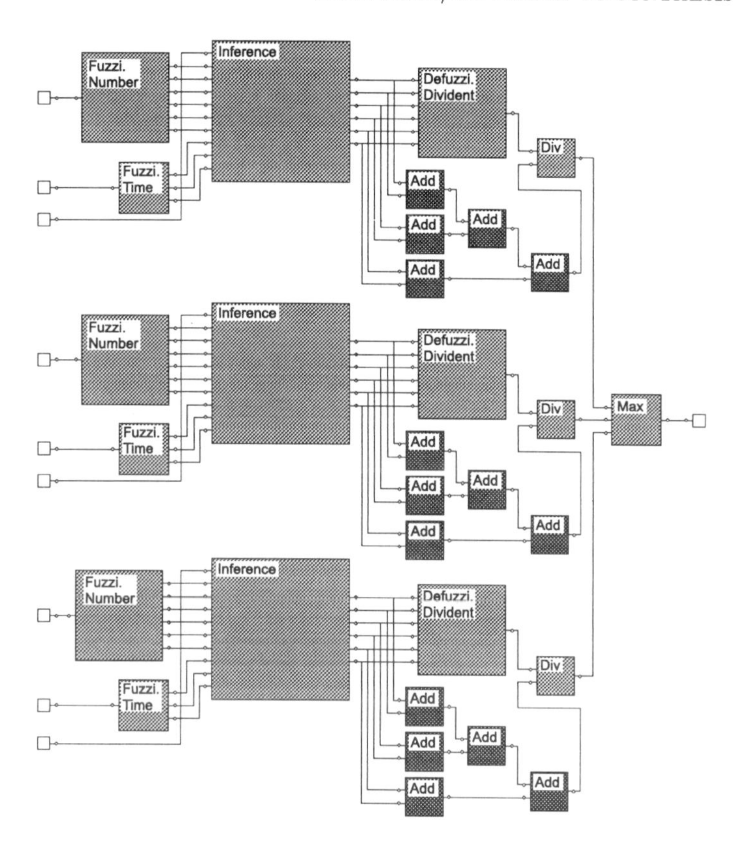

Figure 5.38. Hardware/software partitioning of the fuzzy controller specification

their controllers on the second FPGA. Then, both VHDL descriptions (1266 loc) were synthesized by SYNOPSYS and afterwards the XILINX tools generated the specific configuration files for the FPGAs. The result was a utilization of 80% for the first and 85% for the second FPGA.

After this, the binary file for the DSP was down-loaded to the DAVIS-card using the PC and both FPGAs were configured by down-loading their configuration files from a workstation. At this time, the system could be started using a special simulation environment implemented on the PC generating the values for the inputs and visualizing the results of the fuzzy controller. The correct behavior of the hardware/software implementation was checked with the help of a *logic analyzer* which is very helpful to find errors in the design or on the prototyping board.

Most importantly, this application study proves that the design flow of COOL results in hardware/software systems that are operational and that COOL represents a complete hardware/software co-design tool. A detailed report of the

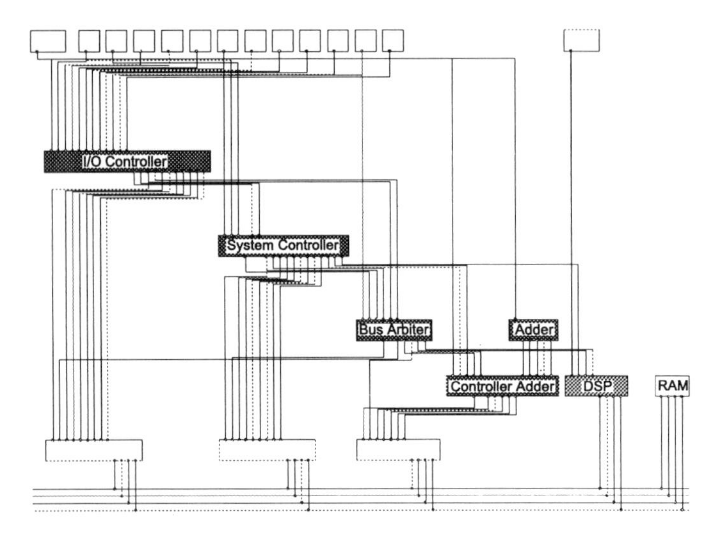

Figure 5.39. Hardware/software netlist of the fuzzy controller

results of this application study (including the complete description of the generated hardware/software systems) is given in [PG293b].

Different alternatives of hardware/software partitions were implemented for the fuzzy controller and in all cases the time to execute the complete design flow from system specification to an implementation on the prototyping environment took not more than 60 minutes. The time-consuming factor was hardware synthesis consuming about 90% of the design time.

Finally, the functionality of a netlist generated by COOL will be described with the help of the partitioned fuzzy controller shown in figure 5.38. The simulated netlist itself is depicted in figure 5.39. In figure 5.40 the results of executing co-simulation with COOL using VANTAGE OPTIUM are depicted. The depicted part of the simulation illustrates the initialization phase and the computation of the first result computed for a set of input values. The main events are enumerated in figure 5.40 and will be described.

Milestones ....

1. The system is reset (`reset <= '0'`) from the environment to be initialized.
2. The processor and the hardware components are initialized.
3. Some inputs change their value, but the system has not been started. Therefore no results are computed.
4. The system is started from the environment (`start <= '1'`).

5. The system controller sends a message to the I/O controller to check if some inputs have changed (`update_inputs <= '1'`).
6. The I/O controller writes input values (which have changed) into memory.
7. The I/O controller indicates to the system controller that input values have changed (`inputs_changed <= '1'`).
8. The system controller writes the execution state of the processor into memory.
9. The processor is started by interrupt (`irq <= '0'`).
10. The processor reads its execution state from memory.
11. The processor executes a set of functions until synchronization with the system controller is required. The computed results are written into memory.
12. Concurrently to step 11, the system controller snoops the bus to detect the DONE-message written by the processor.
13. The system controller has detected the DONE-message of the processor and changes the state-signal of the data path controller to "`state <= 2`" indicating to start execution.
14. In addition, the processor is started again. Therefore, the system controller writes the new state into memory and sets the interrupt afterwards. Remark: now data path and processor work concurrently.
15. The data path controller loads the input values i_1 and i_2 for the data path.
16. The data path is started by the data path controller by `start_dp <= '1'`.
17. The data path computes the result (`o <= 4`) for these inputs and indicates the end of computation to its controller by setting `done_dp <= '1'`.
18. The data path controller writes the computed result into memory.
19. Then, the data path is started again.

   ...

20. The processor is started for computing the last scheduled function by setting the interrupt.
21. The bus snooper of the system controller detects the DONE-message of the processor.
22. The I/O controller reads the computed result (`256`) from memory and transmits it to the environment.
23. Now all steps (starting from step 4) will be repeated if some input signal changes its value.

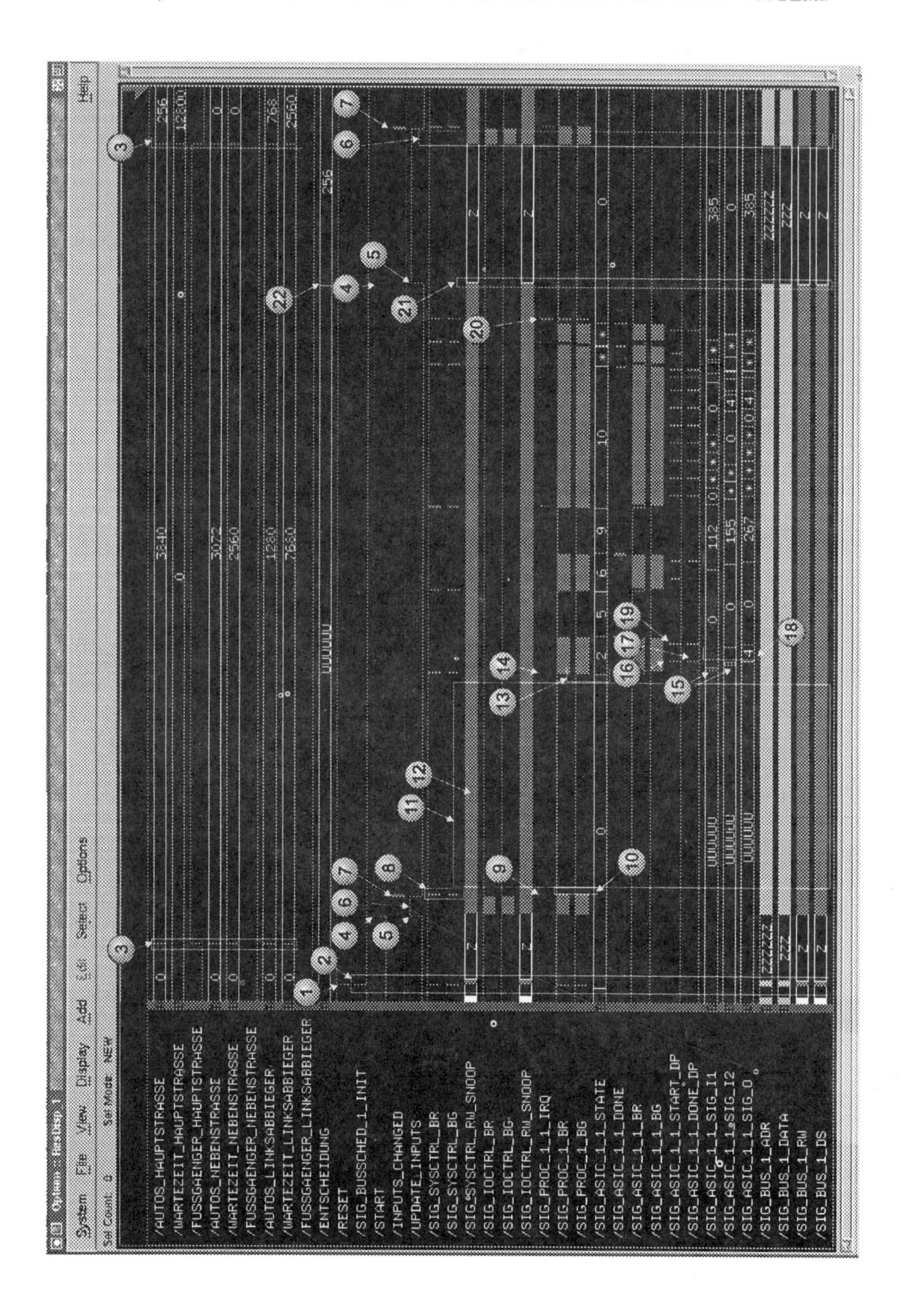

Figure 5.40. Simulating the netlist of the fuzzy controller

6 THE COOL FRAMEWORK

6.1 Implementation of COOL

COOL has been implemented under different UNIX operating systems (SUNOS 4.1 and LINUX) in C++. The total amount of C++ source code is about 130000 lines without considering the set of integrated C and C++ libraries. The following public domain and commercial C/C++ libraries have been used for minimizing the implementation effort required for COOL (current version V1.9):

LEDA: LEDA[1] [Nähe95, MeNa97] is a C++ library of efficient data types and algorithms containing, for example, classes for lists, stacks, queues, graphs and dictionaries. Most of the data types found in the literature are included. The classes for these data types contain most of the necessary algorithms for these data types, simplifying the implementation drastically. Other advantages of LEDA are the following: a) it can be used with almost any C++ compiler and b) the resulting specifications are very easy to read. LEDA is not in the public domain, but can be used freely for academic research and teaching.

WXWINDOWS: WXWINDOWS [Smar96] is a free C++ library for compiling graphical C++ programs on a range of different platforms. WXWINDOWS uses

[1] LEDA: Library of Efficient Datatypes and Algorithms

the native graphical user interface (GUI) on each platform, to be platform-independent. Therefore, programs implemented with WXWINDOWS can be compiled under UNIX (on SUNOS 4.1 or LINUX) using MOTIF, XVIEW and XT. In addition, MS WINDOWS operating systems (WINDOWS 3.1, WINDOWS 95 and WINDOWS NT) are supported using available Windows C++ compilers. WXWINDOWS has been used to implement the graphical user interface of COOL. It has been selected to be able to port COOL to different platforms.

SPI: VTIP [Clsi90] is a commercial VHDL frontend compiling a VHDL specification into an intermediate format called DLS[2] [Clsi91b]. It has been used in COOL for translating the implementation-independent system specification into target-specific

- VHDL specifications for simulation (with VANTAGE OPTIUM) and hardware synthesis (performed by OSCAR and/or SYNOPSYS) and
- C specifications for different processors.

Therefore, the commercial C library SPI[3] [Clsi91a] for handling the DLS data structures has been used to implement the different programs for translating a behavioral VHDL specification.

PGAPACK: To implement the hardware/software partitioning approach based on genetic algorithms, the public domain C library PGAPACK [Levi96] has been used. PGAPACK is a general-purpose, data-structure-neutral, parallel genetic algorithm library. It can easily be used, because it supports the encoding of binary-, integer-, real-, and character-values by alleles. Multiple crossover, mutation and selection operators have been integrated and the genetic optimization process can be steered with a large amount of parameters. The specified programs can be compiled onto uniprocessors, parallel computers, and workstation networks.

COOL is not a stand-alone program but consists of a set of different programs communicating via file exchange formats. Summarizing, 40 programs have been implemented (with a total amount of 40MB disk space under SUNOS 4.1) and about 30 external programs have been interfaced, ranging from very simple ones (such as an editor) to very complex ones (such as a simulator or a synthesis tool).

The graphical user interface of COOL represents a unified design environment, because it is used

1. to specify the inputs (system specification, target architecture and design constraints),

[2]DLS: Design Library System
[3]SPI: Software Procedural Interface

2. to invoke all design steps and
3. to visualize the results with the help of the graphical user interface itself and a set of additional visualization tools, such as
   - graph viewers,
   - waveform display tool,
   - spreadsheet tools generating charts.

   This variety of tools allows a comfortable presentation of results which is a very important aspect in system level design in particular for design space exploration.

The design flow in COOL from system specification to hardware/software implementation is supported by the following tools:

VANTAGE OPTIUM: VANTAGE OPTIUM [Vant94], as a commercial VHDL simulator, has been integrated in the COOL framework for validating the correct functionality of a specified system.

VTIP: The commercial VHDL frontend VTIP has been used to compile the system specification (see SPI described above).

LP_SOLVE, OSL: Hardware/software partitioning based on mixed integer linear programming has been realized with the help of two MILP solvers: the public domain software LP_SOLVE [Berk93] and the commercial solver OSL [OSL92].

OSCAR: High-level synthesis is performed in the COOL framework by OSCAR [LMD94, Land98] for the refined hardware descriptions of the data paths.

SYNOPSYS: After high-level synthesis has been executed, logic synthesis is performed by the commercial logic synthesis tool SYNOPSYS [Syno92]. It is applied to all control-dominated VHDL specifications and the results of OSCAR.

C-COMPILER: Software compilation has been performed by different compilers, such as the public domain compilers for some MOTOROLA DSPs (e.g. [Moto94]).

6.2 Description of COOL

The functionality of COOL should be explained with the help of some snapshots showing some steps during the design process of hardware/software systems with COOL.

6.2.1 Graphical User Interface

COOL includes a graphical user environment, depicted in figure 6.1.

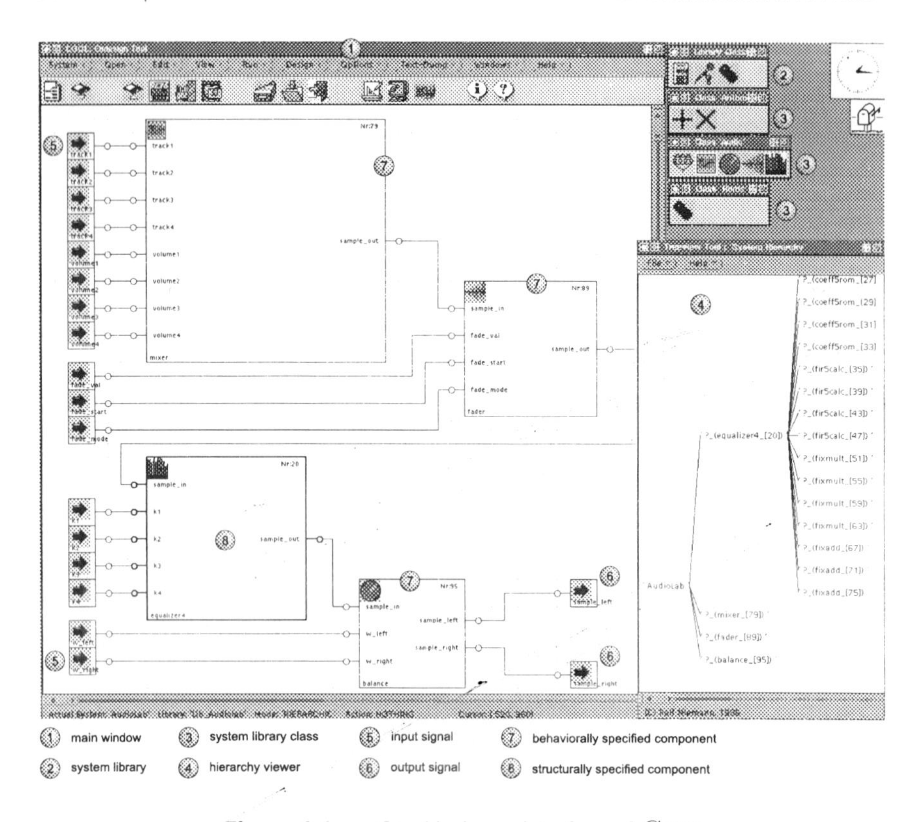

Figure 6.1. Graphical user interface of COOL

The main window (1) is used for specifying a system graphically by instantiating pre-defined components. These pre-defined components are stored in classes (3) of the system library (2). Using the drag-and-drop technique, systems can be specified. Each specified system can then also be stored in the system library. Therefore, the specification methodology can be either top-down or bottom-up and the resulting systems may be hierarchical. For this reason, the designer can go down the hierarchy from the top to the bottom level. In addition, a hierarchy viewer (4) has been implemented to give the designer an overview of the structure of the system. In the example given, an audio application is depicted. It consists of several input ports (5), two output ports (6) and four instances of pre-defined system components. Three of these components represent behavioral VHDL descriptions (7), one has been structurally specified (8) illustrated by the bold outline.

6.2.2 Validation using Simulation

To validate the correct functionality of a system, the commercial VHDL simulator VANTAGE OPTIUM has been integrated in COOL, as described in section 3.5. In figure 6.2, an audio **fader** has been specified by a behavioral **VHDL** description (1).

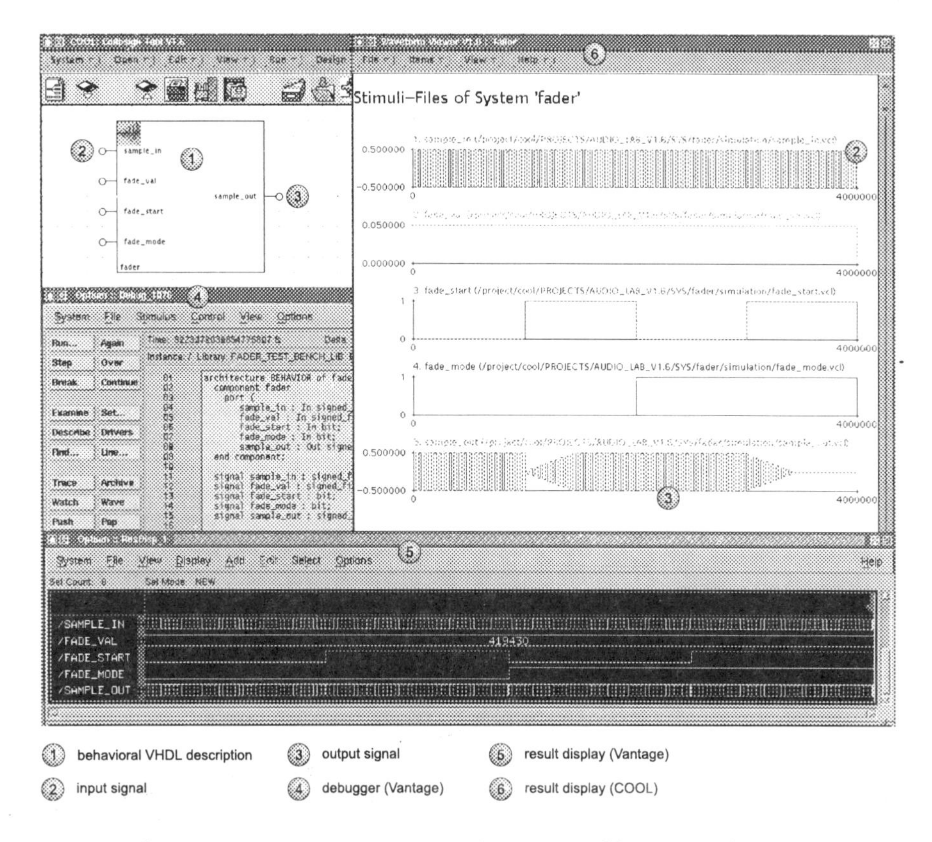

Figure 6.2. Simulation of Specification with VANTAGE OPTIUM

The user has defined the stimuli files for the input signals (2) and some result files for the output signals (3). After the simulation has been prepared, the VANTAGE OPTIUM simulator (4) is started. After the simulation has finished, the result display (5) of VANTAGE OPTIUM displays the resulting waveforms for all signals. One problem is that some values are fixed-point numbers, being a subtype of bit vectors in COOL. Therefore, it is difficult to check the results with the results display of VANTAGE OPTIUM, because data can only be displayed as binary, octal, decimal or hexadecimal numbers. The result viewer (6) integrated in COOL allows a comfortable visualization possibility, because fixed-point numbers are supported and displayed as x-y-plots. In this example, **sample_in** represents an alternating signal. **fade_value** is constant 0.05 defining that incoming samples are faded in or out with 5% reduction. First, the incoming samples are faded in, and afterwards the samples are faded out. Therefore, the waveform of the output signal **sample_out** (3) is correct.

6.2.3 Specification of Design Constraints

A great advantage of CooL is that the graphical user environment provides a unified design environment. In addition to the specification of the system, the design constraints are also specified graphically. The possibility of specifying a variety of design constraints is very important to exploit the experience of the designer. In figure 6.3, some examples of design constraints supported in CooL are given.

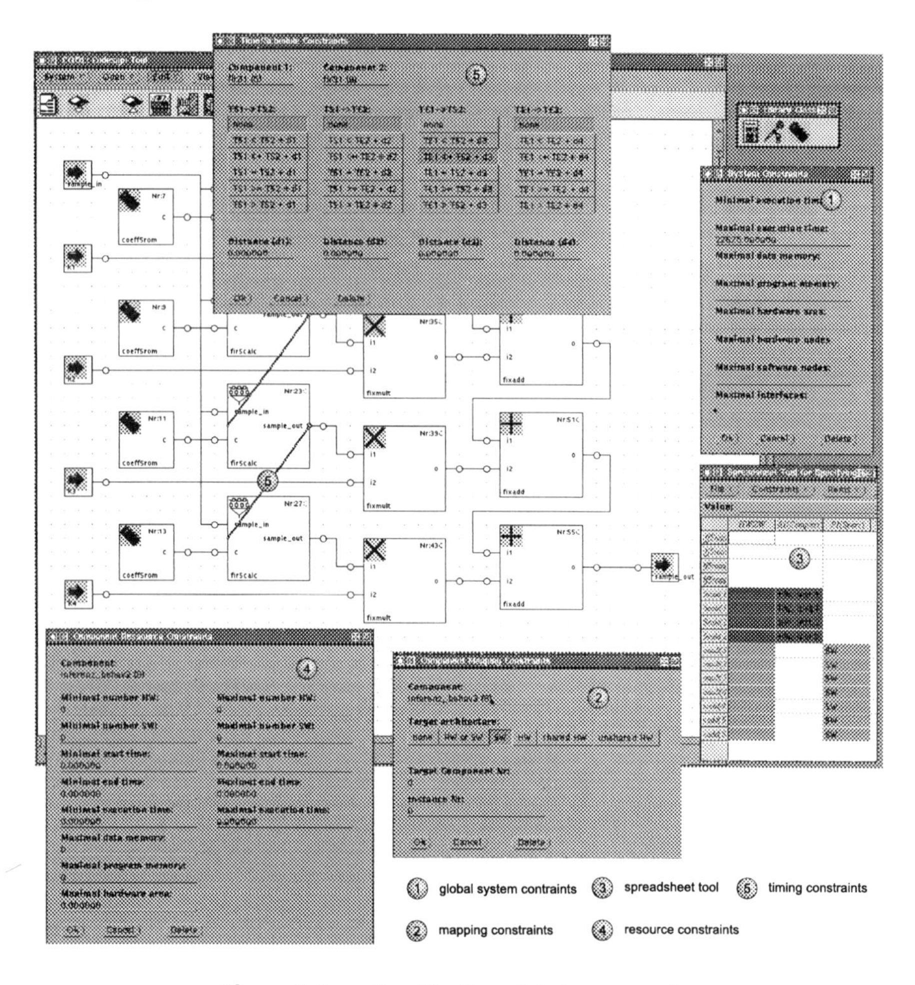

Figure 6.3. Specification of design constraints

First, the designer is able to define global timing and resource constraints (1) related to the overall design costs (e.g. execution time, hardware area, memory usage). In addition, he can determine mapping and resource constraints. These constraints can be defined for single instances of components, for all instances of a certain component and for each group of instances. The definition of these mapping/resource constraints is done with the help of dialog (2)/(4) for each constraint or an integrated spreadsheet tool for all constraints (3) to give a better overview. Relative timing constraints between two instances can also be defined very efficiently via the dialog (5) or the spreadsheet tool (3).

6.2.4 Hardware/Software Partitioning

After the system has been partitioned, the designer can store the resulting design in a design library. The partitioning approach based on genetic algorithms has the advantage that not only one but a set of different solutions is computed.

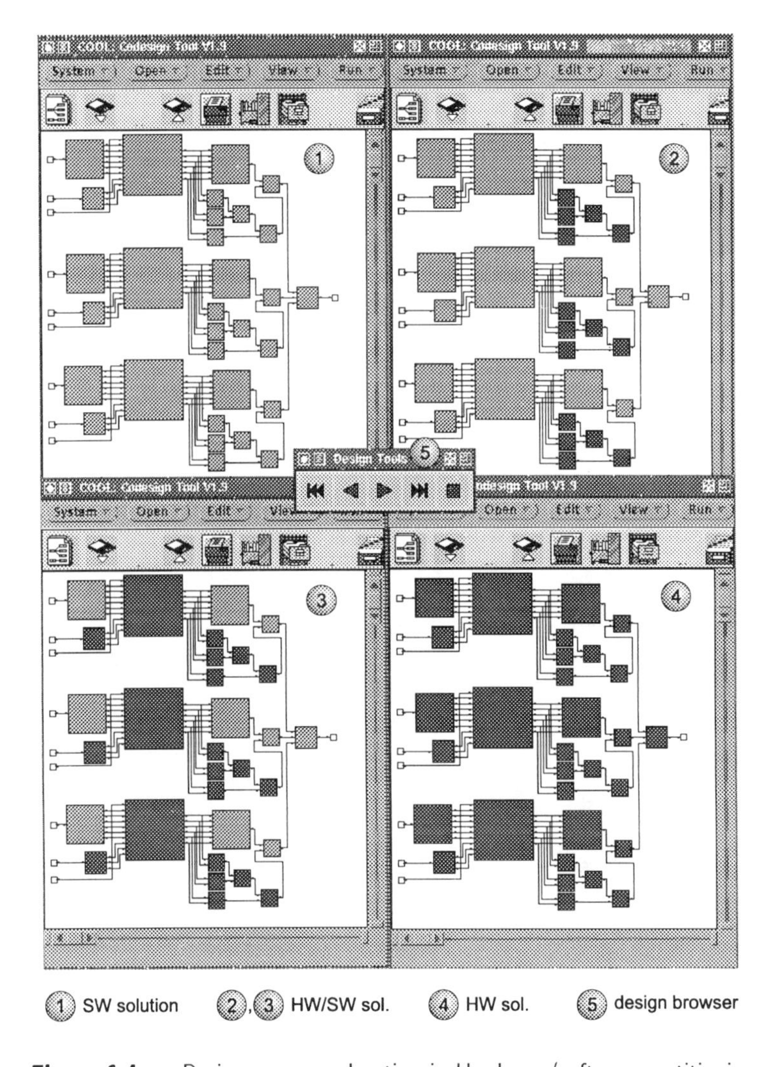

Figure 6.4. Design space exploration in Hardware/software partitioning

Therefore, the designer should be able to compare these different implementations of the system, as depicted in figure 6.4. For this purpose, a design viewer (5) has been integrated in the graphical user environment of COOL with which the designer is able to explore the design space. For each solution the structural system specification is colored. The user can iteratively load all designs stored in the design library and can get detailed information for each design, such as the design costs, the resulting schedule and a lot of statistics. Thus, the designer is able to select the best design corresponding to his personal preferences. In figure 6.4, four different partitions of the fuzzy system are depicted, ranging from a pure software solution (1) and mixed hardware/software implementations (2),(3) to a pure hardware (4).

6.2.5 Co-Synthesis

After hardware/software partitioning, COOL generates a netlist of the complete hardware/software system during co-synthesis. In figure 6.5, the corresponding netlists for the partitions shown in figure 6.4 are depicted.

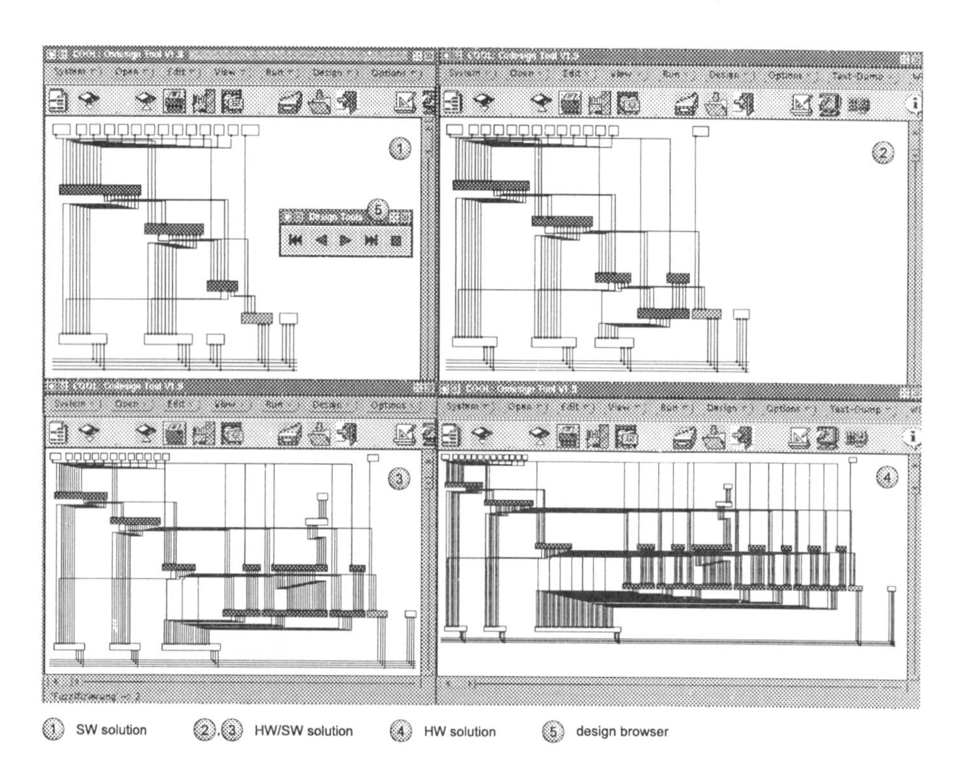

Figure 6.5. Design space exploration in co-synthesis

6.2.6 *Co-Simulation*

Finally, the generated netlist can be simulated with the help of the interface to VANTAGE OPTIUM. Thus, the designer is able to compare the results of simulating the implementation-independent system specification with the results of co-simulating the generated netlist. In figure 6.6, both the results of simulating the system specification and the generated netlist of the fuzzy system are depicted.

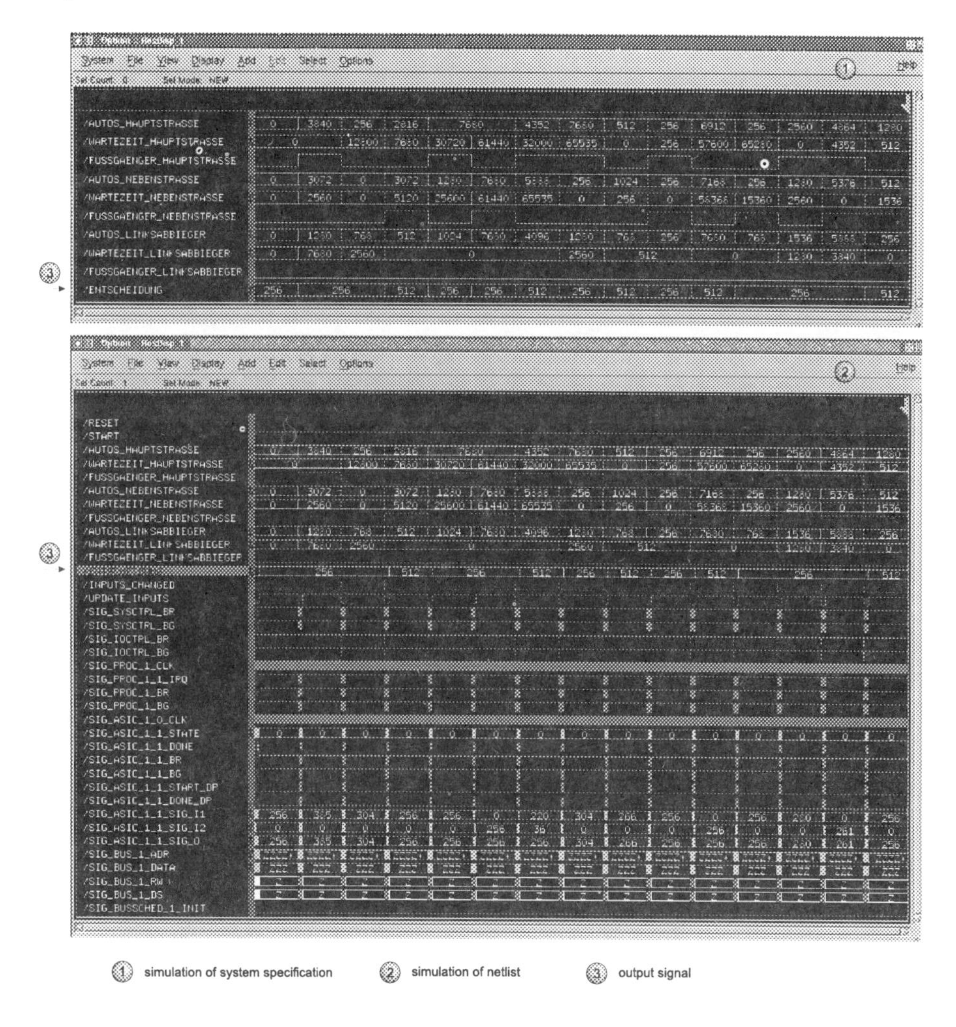

Figure 6.6. Simulation of system specification and netlist

Clearly, the simulated waveforms for the output signal (3) are identical for simulating the system specification (1) and the refined hardware/software implementation (2). The main difference is that if an input signal changes its value, the output signal simultaneously changes without any delay when simulating the system specification. The reason for this is that the system specification does not contain any timing construct. In contrast, when simulating the netlist a delay occurs between the points of time where input and output values change. The reason for this is that the refined hardware/software specifications contain timing information, e.g. the hardware description works synchronously to the clock. However, the results for the output signals are the same.

7 SUMMARY AND CONCLUSIONS

This book has presented an approach to *hardware/software co-design* for data flow dominated systems. It represents the attempt of bridging the gap between hardware and software design for *embedded systems*. This chapter will summarize and evaluate the results which have been presented. After this, an overview of future work will conclude this book.

7.1 Summary and Contribution to Research

In this book, first the existing approaches in the area of hardware/software co-design have been summarized and compared. Then, a new hardware/software co-design tool, called COOL, has been presented, a tool which is able to implement an abstract system specification on a *heterogeneous target architecture*. Main research focus has been put on the heterogeneous implementation problems of these applications: *hardware/software partitioning* and *co-synthesis*. In contrast to most co-design approaches, the target architecture is not restricted in the number of ASICs and processors. COOL supports the design process for target architectures consisting of multiple processors, multiple ASICs and a set of communication channels connecting these processing units. Besides SPECSYN, COOL is the only hardware/software co-design tool automating the hardware/software partitioning and co-synthesis phase for this general type of target architectures. All other co-design approaches restrict the target architecture, e.g. single-processor architectures, or the design steps are not fully automated, e.g. manual partitioning. The strength of COOL is the high degree of automation while considering user-defined constraints. The support of user-defined constraints is very important, because the valuable experience of designers has to be exploited.

System Specification:. The types of applications supported by COOL are mainly data flow dominated systems with hard timing constraints, such as audio and video applications. In COOL, systems are specified in a homogeneous way using a subset of the hardware description language VHDL which presently represents the most popular language for the design of digital systems. The systems are specified using behavioral or structural VHDL resulting in hierarchical specifications. In addition to the system specification, the designer has to define the target architecture and the design constraints.

Cost Estimation:. After a system has been specified, for all system components which have been specified using behavioral VHDL, costs for hardware and software implementations are estimated. In contrast to nearly all approaches in the literature, this is done by using the tools of the design process instead of using special estimation tools. This means, for each processor of the target architecture COOL generates a C specification, compiles it with the corresponding compiler and analyzes the generated assembly code afterwards. On the hardware side, a synthesizable hardware specification is generated which is synthesized by the high-level synthesis tool OSCAR. *High-level synthesis* represents the ideal interface between system level design and logic synthesis, because of the trade-off between computation time and accuracy. After this, the generated control-step list of OSCAR is analyzed. *Static analysis* methods can be performed during cost estimation, because COOL restricts itself to systems specified without non-deterministic behaviors, such as unbounded loops. The reason for this is that COOL supports data flow dominated systems with hard timing constraints. Therefore, the worst-case execution time has to be estimated. The results of the cost estimation step are high-precision values for the hardware and software cost metrics. Hardware cost metrics considered by COOL are execution time and chip area, software cost metrics include execution time and program/data memory usage.

Hardware/Software Partitioning:. The presented cost metrics are used in hardware/software partitioning for computing a solution for a certain objective function, in most cases hardware minimization under timing constraints. Hardware/software partitioning is solved in COOL very precisely, because the most important aspects are taken into account, e.g. mapping, scheduling, hardware sharing, interfacing and functional pipelining constraints. In particular, the way of modelling additional interface costs is very precise compared to related work found in the literature. Different algorithms have been presented to solve the hardware/software partitioning problem. The first approach formulates the hardware/software partitioning problem as a *mixed integer linear programming problem.* Thus, the computed results are optimal, but only problems of small complexity can be handled, because the hardware/software partitioning problem is NP-hard. To handle larger designs, a second algorithm has been developed separating the mapping and the scheduling aspects of the problem into two phases. This *heuristic scheduling* approach drastically reduces the computation time while computing nearly optimal results. Nevertheless, large problems can not be solved efficiently in some cases. In addition, both approaches suffer from the fact that computation time is not predictable and that only one solution is computed. For this reason, a third partitioning algorithm has been developed based on *genetic algorithms.* The advantages of this approach are manifold: First, the computation time is predictable (it depends on the number of generations and the selected population size) and a set of solutions is computed. This is a very important aspect, because design space exploration is of great interest in co-design. It has been shown by extensive benchmarking that the results obtained by the genetic algorithm approach are nearly optimal. Furthermore, the genetic algorithm approach has been adapted to solve the more general *extended partitioning problem.* This problem considers not only a single hardware implementation for each system component, but a complete *AT-curve* representing a set of implementation alternatives. Genetic algorithms are very flexible, and therefore other extensions to the partitioning problem can easily be integrated.

Co-synthesis:. Most approaches found in the literature work either on hardware/software partitioning or co-synthesis. However, COOL automates the solution to both problems. After the partitioning has computed a hardware/software mapping and a schedule, the co-synthesis phase generates a heterogeneous implementation based on distributed communicating controllers. Algorithms based on a state-transition graph have been presented which automate this step. The result of the co-synthesis phase in COOL is a set of target-specific descriptions for processors and ASICs, a set of additional hardware components (e.g. a system controller, an I/O controller, data path controllers and bus arbiters) and a netlist wiring all ASICs, processors and additional components. This generated netlist serves as an input either to co-simulation tools or design tools for synthesizing hardware and compiling software in COOL.

Design Methodology Management:. One strength of COOL is that it represents a unified design environment. First, the user specifies the inputs (system specification, design constraints, target architectures) using the graphical user environment. Then, he invokes the necessary design steps from the same environment. After the tools have computed one design or a set of designs, the results are visualized and the designer can very easily choose a certain implementation.

7.2 Future Work

COOL represents a complete hardware/software co-design tool capable of implementing a specification on a heterogeneous multi-processor-multi-ASIC target architecture. Nevertheless, there is scope for further development.

System Specification:. COOL uses a subset of VHDL to specify a system. For our purposes (research in hardware/software partitioning and co-synthesis), this approach was sufficient. But to commercialize a tool like COOL most of remaining work is related to system specification.

Cost Estimation:. New cost metrics could be considered during partitioning, for example, *low power* or *testability metrics*. In such cases, new estimation tools have to be developed for these new cost metrics.

Hardware/Software Partitioning:. Hardware/software partitioning can be extended by different aspects:

- One possible extension is to optimize not only the implementation of a specification on a fixed target architecture during hardware/software partitioning, but also the target architecture itself. In such a case, the target architecture would be defined as a set of processors and ASICs being potential candidates for implementing parts of the system.
- A second extension is a partitioning approach based on variable granularity of the system specification. Very often it is difficult to determine the granularity of system components which should be mapped either to software and hardware. If the system is specified too fine-grained, solving the partitioning problem may be too time-consuming. If it is too coarse-grained, it may result in too expensive designs. For this reason, a partitioning approach considering different levels of granularity would be a good solution to this problem.

Co-synthesis:. Co-synthesis is solved in COOL using a fixed communication scheme, based on additional hardware implementing a distributed run-time scheduler. The implemented protocols are low-level handshake protocols, the communication media are busses and dedicated lines. The presented partitioning approach of COOL is flexible enough to incorporate communication synthesis to allocate a certain communication mechanism and the corresponding protocol. Therefore most of the work in co-synthesis is related to extending the flexibility of the implemented communication scheme. Therefore, the following aspects will be considered in the future:

- A library of existing communication units and corresponding protocols would allow to realize communication synthesis during hardware/software partitioning. Using this library, the communication between different processing units could be optimized.

- Communication steered by software processors will be examined to minimize the additinal amount of hardware.
- Design time can be shortened when reusing existing peripherals or devices in some cases instead of synthesizing new ASICs. Some devices have fixed protocols and for this reason additional *interface synthesis* algorithms such as *protocol conversion* would be required to instantiate these components into the design.

In addition to the optimization of the communication, another important step is the integration of ASIPs. In industry, e.g. in consumer electronics with high-volume products, the usage of ASIPs is of major interest to reduce the power consumption and the production costs. To automate the design of heterogeneous systems using ASIPs, *retargetable code generators* are necessary. Therefore, integrating such a compiler (e.g. RECORD [Leup97]) in COOL is of great interest.

Co-simulation:. To validate the correct functionality of the generated hardware/software systems, co-simulation is executed. Co-simulation approaches on a workstation always suffer from the fact that simulation speed is very low. For this reason, prototyping boards have gained more and more importance. These boards are used for implementing the hardware parts of the system with the help of re-programmable hardware units, e.g. FPGAs. These boards are then connected to the instantiated processors for simulation. One key problem of this approach is that the hardware specification has to be partitioned onto a set of FPGAs, considering CLB, pin and timing requirements.

Summarizing, the tasks presented above represent only some of the aspects which can be considered in future. For some of these problems preliminary solutions have been published, but there is still a lot of remaining work to be done to come up with a commercial hardware/software co-design tool.

References

[Ashe89] P.J. Ashenden. *The VHDL Cookbook*. University of Adelaide, First Edition, 1989.

[Axel97] J. Axelsson. *Architecture Synthesis and Partitioning of Real-Time Systems: A Comparison of Three Heuristic Search Strategies*. International Workshop on Hardware/Software Codesign, pp. 161–165, 1997.

[BCO95] G. Borriello, P. Chou and R.B. Ortega. *Embedded System Co-Design: Towards Portability and Rapid Integration*. Hardware/Software Co-Design, M.G. Sami and G. De Micheli, Eds., Kluwer Academic Publishers, pp. 243–264, 1995.

[BeGo92] G. Berry and G. Gonthier. *The Esterel Synchronous Programming Language: Design, Semantics, Implementation*. Science of Computer Programming vol. 19, no. 2, pp. 87-152, 1992.

[Bend96] A. Bender. *Design of an Optimal Loosely Coupled Heterogeneous Multiprocessor System*. In Proceedings of the European Design & Test Conference (ED&TC), pp. 275–281, 1996.

[Berg95] J.-M. Berge. *High-Level System Modelling*. Kluwer Academic Publishers, 1995.

[Berk93] M.R.C.M. Berkelaar. *Manual Page of lp_solve*. `ftp://ftp.es.ele.tue.nl/pub/lp_solve/`, Eindhoven University of Technology, Design Automation Section, 1992.

[Berr92] G. Berry. *Esterel on Hardware*. Mechanized Reasoning and Hardware Design, pp. 87-104, 1992.

[BFS93] A. Balboni, W. Fornaciari and D. Sciuto. *Partitioning and Exploration Strategies in the TOSCA Co-Design Flow*. International Workshop on Hardware/Software Codesign96, pp. 62–69, 1996.

[BFS95] A. Balboni, W. Fornaciari and D. Sciuto. *TOSCA: A Pragmatic Approach to Co-Design Automation of Control-dominated Systems.* Hardware/Software Co-Design, M.G. Sami and G. De Micheli, Eds., Kluwer Academic Publishers, pp. 265–294, 1995.

[BFS96] A. Balboni, W. Fornaciari and D. Sciuto. *Co-synthesis and Co-simulation of Control-dominated Embedded Systems.* Design Automation for Embedded Systems, vol. 1, no. 3, pp. 257–289, 1996.

[BHLM91] J. Buck, S. Ha, E.A. Lee and D.G. Messerschmitt. *Ptolemy: A Platform for Heterogeneous Simulation and Prototyping.* European Simulation Conference, 1991.

[BHS91] F. Belina, D. Hogrefe and A. Sarma. *SDL With Applications From Protocol Specification.* Prentice Hall, 1991.

[Blic97] T. Blickle. *Theory of Evolutionary Algorithms and Application to System Synthesis.* A dissertation submitted to the Swiss Federal Institute of Technology Zurich, Diss. ETH No. 11894, 1997.

[BMM98] A. Basu, R.S. Mitra and P. Marwedel. *Interface Synthesis for Embedded Applications in Codesign Environment.* In Proceedings of the International Conference on VLSI Design, 1998.

[Booc91] G. Booch. *Object-oriented Design with Applications.* Benjamin / Cummings, Redwood City, California, 1991.

[BRX94] E. Barros, W. Rosenstiel and X. Xiong. *A method for partitioning UNITY language in hardware and software.* In Proceedings of the European Design Automation Conference (EURO-DAC), pp. 220–225, 1994.

[BuVe92] K. Buchenrieder and C. Veith. *Codes: A practical concurrent design environment.* International Workshop on Hardware/Software Codesign, 1992.

[Calv93] J.P. Calvez. *Embedded Real-Time Systems.* Wiley Professional Computing, 1993.

[Calv96] J.P. Calvez. *A CoDesign Case Study with the MCSE Methodology.* Design Automation for Embedded Systems, vol. 1, no. 3, pp. 183–212, 1996.

[CaWi96] R. Camposano and J. Wilberg. *Embedded System Design.* Design Automation for Embedded Systems, vol. 1, no. 1-2, pp. 5–50, 1996.

[CEG+96] M. Chiodo, D. Engels, P. Giusto, H. Hsieh, A. Jurecska, L. Lavagno, K. Suzuki, A. Sangiovanni-Vincentelli. *A Case Study in Computer-Aided Co-design of Embedded Controllers.* Design Automation for Embedded Systems, vol. 1, no. 1-2, pp. 51–67, 1996.

[Chen77] P.S. Chen. *The Entity-Relationship Approach to Logical Data Base Design.* Q.E.D. Information Sciences, Wellesley, Massachusetts, 1977.

[CKL96] W.-T. Chang, A. Kalavade and E.A. Lee. *Effective Heterogeneous Design and Co-Simulation.* Hardware/Software Co-Design, M.G. Sami and G. De Micheli, Eds., Kluwer Academic Publishers, pp. 187–213, 1995.

[CLL+96] C. Carreras, J.C. López, M.L. López and C. Delgado-Kloos. *A Co-Design Methodology Based on Formal Specification and High-level Estimation.* International Workshop on Hardware/Software Codesign, pp. 28-35, 1996.

[Clsi90] CAD Language Systems Inc.. *VTIP: VHDL Tool Integration Platform, User Manual.* 1990.

[Clsi91a] CAD Language Systems Inc.. *Software Procedural Intercvace (SPI), User Manual.* 1991.

[Clsi91b] CAD Language Systems Inc.. *Design Library System (DLS) - The VHDL View, User Manual.* 1991.

[COB92] P. Chou, R.B. Ortega and G. Boriello. *Synthesis of the HW/SW Interface in Microcontroller-Based Systems.* In Proceedings of the International Conference on Computer-Aided Design (ICCAD), pp. 488–495, 1992.

[COB95a] P. Chou, R.B. Ortega and G. Borriello. *The Chinook Hardware/Software Co-Synthesis System.* In Proceedings of the International Symposium on System Synthesis (ISSS), pp. 22–27, 1995.

[COB95b] P. Chou, R.B. Ortega and G. Boriello. *Interface Co-Synthesis Techniques for Embedded Systems.* In Proceedings of the International Conference on Computer-Aided Design (ICCAD), pp. 280–287, 1995.

[Coel89] D.R. Coelho. *The VHDL Handbook.* Kluwer Academic Publishers, Boston, 1989.

[Cohe81] D. Cohen. *On holy wars and a plea for peace.* Computer 14:10 (October), pp. 48–54, 1981.

[Comp91] Compass Design Automation. *Users Manual Version V8R3.* 1991.

[Davi93] DigAs (Digital Audio Systems). *DAVIS Digital Audio Virtual Instrument System, Desktop and DAVIS DSP Board.* Technical Documentation, 1993.

[DFM+97] J.M. Daveau, G.F. Marchioro, C.A. Valderama and A.A. Jerraya. *VHDL generation from SDL specifications.* Hardware Description

Languages and their Applications, pp. 182-201, Chapman & Hall, 1997.

[DIJ95] J.M. Daveau, T.B. Ismail and A.A. Jerraya. *Synthesis of System-Level Communication by an Allocation-Based Approach.* In Proceedings of the International Symposium on System Synthesis (ISSS), pp. 150–155, 1995.

[DMIJ97] J.M. Daveau, G.F. Marchioro, T.B. Ismail and A.A. Jerraya. *Protocol Selection and Interface Generation for HW-SW Codesign.* IEEE Transactions on VLSI Systems, Vol. 5, No. 1, pp. 136–144, 1997.

[Döme94] R. Dömer. *Optimale Mikroarchitektursynthese mittels Ganzzahliger Programmierung.* Diploma Thesis, University of Dortmund, 1994.

[Döme96] K. Dömer. *Spezifikation und Hardware/Software Kostenabschätzung eines MPEG-Audio Systems.* Diploma Thesis, University of Dortmund, 1996.

[Ecke93] W. Ecker. *Using VHDL for HW/SW Co-Specification.* Corporate Research and Development, Siemens AG Munich, Germany, 1993.

[EHB93] R. Ernst, J. Henkel and T. Benner. *Hardware/Software Co-Synthesis for Microcontrollers.* IEEE Design and Test of Computers, Vol. 10 No. 4, pp. 64-75, Dec. 1993.

[EPD94] P. Eles, Z. Peng and A. Doboli. *VHDL System-level Specification and Partitioning in a Hardware/Software Co-Synthesis Environment.* International Workshop on Hardware/Software Codesign, pp. 49–55, 1994.

[EPKD97] P. Eles, Z. Peng, K. Kuchinski and A. Doboli. *System Level Hardware/Software Partitioning Based on Simulated Annealing and Tabu Search.* Design Automation for Embedded Systems, vol. 2, no. 1, pp. 5–32, 1997.

[EKP98] P. Eles, K. Kuchinski and Z. Peng. *System Synthesis with VHDL.* Kluwer Academic Publishers, 1998.

[Esse96] R. Esser. *An Object Oriented Petri Net Approach to Embedded System Design.* Dissertation ETH Zurich, Diss. ETH No. 11869, 1996.

[EVD89] P. H. J. van Eijk, C.A. Vissers and M. Diaz. *The formal description technique LOTOS.* Elsevier Science Publishers B.V., 1989.

[Frit96] C. Fritsch. *Hardware/Software Cosynthese eines MPEG-Audio Systems.* Diploma Thesis, University of Dortmund, 1996.

[GaJo79] M.R. Garey and D.S. Johnson. *Computers and Intractability, A Guide to the Theory of NP-Completeness.* Freeman and Company, New York, 1979.

[GaKu83] D. Gajski and R.H. Kuhn. *New VLSI Tools.* IEEE Computer, pp. 11–14, 1987.

[GaVa95] D. Gajski and F. Vahid. *Specification and Design of Embedded Hardware-Software Systems.* IEEE Design and Test of Computers, pp. 53-67, Spring 1995.

[GCM92] R.K. Gupta, C. Coelho and G. De Micheli. *Synthesis and Simulation of Digital Systems Containing Interacting Hardware and Software Components.* In Proceedings of the Design Automation Conference (DAC), pp. 225–230, 1992.

[GGB96] J. Gong, D. Gajski and S. Bakshi. *Model Refinement for Hardware-Software Codesign.* In Proceedings of the European Design & Test Conference (ED&TC), pp. 270–274, 1996.

[GGB97] J. Gong, D. Gajski and S. Baksi. *Model Refinement for Hardware-Software Codesign.* ACM Transactions on Design Automation of Electronic Systems, Vol. 2, No. 1, January 1997, pp. 22–41.

[Gold89] D.E. Goldberg. *Genetic Algorithms in Search, Optimization and Machine Learning.* Addison-Wesley Publishing Company, Inc., 1989.

[Gomo60] R.E. Gomory. *Solving Linear Programming Problems in Integers.* Combinatorial Analysis, R. E. Bellman, M. Hall (Eds.), American Mathematical Society, pp. 211–216, 1960.

[GuLi97] R.K. Gupta and S.Y. Liao. *Using a Programming Language for Digital System Design.* IEEE Design and Test of Computers, pp. 72-80, 1997.

[GuMi90] R.K. Gupta and G. De Micheli. *Partitioning of Functional Models of Synchronous Digital Systems.* In Proceedings of the International Conference on Computer-Aided Design (ICCAD), pp. 216–219, 1990.

[GuMi92] R.K. Gupta and G. De Micheli. *System-Level Synthesis Using Re-programmable Components.* In Proceedings of the European Design Automation Conference (EURO-DAC), pp. 2–7, 1992.

[GuMi93] R.K. Gupta and G. De Micheli. *Hardware/Software Cosynthesis for Digital Systems.* IEEE Design and Test of Computers, pp. 29–49, 1993.

[GuMi96] R.K. Gupta and G. De Micheli. *A Co-Synthesis Approach to Embedded System Design Automation.* Design Automation for Embedded Systems, vol. 1, no. 1-2, pp. 69–120, 1996.

[Gupt94] R.K. Gupta. *Co-Synthesis of Hardware and Software for Digital Embedded Systems.* PhD Thesis, 1994.

[GVN94] D. Gajski, F. Vahid and S. Narayan. *A System-Design Methodology: Executable-Specification Refinement.* In Proceedings of the European Design & Test Conference (ED&TC), pp. 458–463, 1994.

[GVNG94] D. Gajski, F. Vahid, S. Narayan and J. Gong. *Specification and Design of Embedded Systems.* Prentice Hall, 1994.

[GVNG96] D. Gajski, F. Vahid, S. Narayan and J. Gong. *SpecSyn: AN Environment Supporting the Specify-Explore-Refine Paradigm for Hardware/Software System Design.* Technical Report CS-96-08, University of California, Irvine, 1996.

[HaPn88] D. Harel, A. Pneuli et. al. *STATEMATE: A working environment for the development of complex reactive systems.* Proceedings 10th International Conference on Software Engineering, IEEE Press, pp. 396–406, New York, 1988

[Hare87] D. Harel. *Statecharts: A visual formalism for complex systems.* Science of Computer Programming 8, 1987.

[HEY+95] J. Henkel, R. Ernst, W. Ye, M. Trawny and T. Benner. *COSYMA: Ein System zur Hardware/Software Co-Synthese.* GME Fachbericht Nr. 15 Mikroelektronik, pp. 167–172, 1995.

[HHE94] D. Henkel, J. Herrmann and R. Ernst. *An Approach to the Adaption of Estimated Cost Parameters in the COSYMA System.* International Workshop on Hardware/Software Codesign, pp. 100–107, 1994.

[Hilf85] P. Hilfinger. *A high-level language and silicon compiler for digital signal processing.* In Proceedings of the Custom Integrated Circuits Conference, 1985.

[Hoar78] C.A.R. Hoare. *Communicating sequential processes.* Communications of the ACMs, 21(8): pp. 666-677, 1978.

[Hogr89] D. Hogrefe. *Estelle, LOTOS und SDL.* Springer Verlag Berlin, 1989.

[Holl75] J.H. Holland. *Adaption in Natural and Artificial Systems.* The University of Michigan Press, Ann Arbor, MI, 1975.

[HoUl79] J.E. Hopcroft and J.D. Ullman. *Einführung in die Automatentheorie, Formale Sprachen und Komplexitätstheorie.* Addison-Wesley, 1979.

[IEEE87] IEEE. *IEEE Standard VHDL Language Reference Manual. IEEE Std. 1076-1987, IEEE Inc.* New York, 1987.

[IEEE93] IEEE. *IEEE Standard VHDL Language Reference Manual. IEEE Std. 1076-1993 (Revision of IEEE Std. 1076-1987).* IEEE Inc., New York, 1993.

[IEEE96] IEEE. *Hardware Description Language Based on the Verilog(TM) Hardware Description Language. IEEE Std. 1364-1996.* IEEE Inc., 1996.

[IKJ94] T.B. Ismail, K.O'Brien and A.A. Jerraya. *Interactive System-Level Partitioning with PARTIF.* In Proceedings of the European Conference on Design Automation (EDAC), pp. 464–468, 1994.

[Inmo84] Inmos Ltd.: . *OCCAM Programming Manual.* Prentice Hall, 1984.

[IsJe95] T.B. Ismail and A.A. Jerraya. *Synthesis Steps and Design Models for Codesign.* IEEE Computer, pp. 44–52, 1995.

[ISO87] ISO Standard ISO/DIS 9074. *Estelle (Formal Description Technique Based on an Extended State Transition Model).* 1987.

[ISO89] ISO Standard IS 8807. *LOTOS a Formal Description Technique Based on the Temporal Ordering of Observational Behavior.* 1989.

[ITU88] ITU. *Recommendation Z.100: Specification and Description Language SDL, volume X.1-X.5.* ITU, 1988.

[ITU92] ITU. *Recommendation Z.100: Specification and Description Language SDL, volume X.R25-X.R32.* ITU, 1992.

[Jens90] K. Jensen. *Coloured Petri Nets: A High Level Language for System Design and Analysis.* Advances in Petri Nets 1990, G. Rozenberg, LNCS 483, Springer-Verlag, 1990.

[JEO+94] A. Jantsch, P. Ellervee, J. Öberg, A. Hemani and H. Tenhunen. *Hardware/Software Partitioning and Minimizing Memory Interface Traffic.* In Proceedings of the European Design Automation Conference (EURO-DAC), pp. 226–231, 1994.

[JeOb94] A.A. Jerraya and K. O'Brien. *SOLAR: An Intermediate Format for Hardware for System-Level Modelling and Synthesis.* in Computer Aided Software/Hardware Engineering, J. Rozenblit, K. Buchenrieder, Eds., IEEE Press 1994.

[KAJW93] S. Kumar, J.H. Aylor, B.W. Johnson and W.A. Wulf *The Codesign of Embedded Systems - A Unified Hardware Software Representation.* Kluwer Academic Publishers, 1993.

[KaLe93] A. Kalavade and E.A. Lee. *A Hardware-Software Codesign Methodology for DSP Applications.* IEEE Design and Test of Computers, pp. 16–28, 1993.

[KaLe94] A. Kalavade and E.A. Lee. *A Global Critically/Local Phase Driven Algorithm for the Constrained Hardware/Software Partitioning Problem.* International Workshop on Hardware/Software Codesign, pp. 42–48, 1994.

[KaLe95] A. Kalavade and E.A. Lee. *The Extended Partitioning Problem: Hardware/Software Mapping and Implementation-Bin Selection.* International Workshop on Rapid Systems Prototyping, 1995.

[KaLe97] A. Kalavade and E.A. Lee. *The Extended Partitioning Problem: Hardware/Software Mapping, Scheduling and Implementation-Bin Selection.* Design Automation for Embedded Systems, vol. 2, no. 2, pp. 125–163, 1997.

[Karm84] N. Karmakar. *A new Polynomial Algorithm for Linear Programming.* Combinatorica, 4 (4), pp. 373–395, 1984.

[KeRi78] B. Kernighan and D. Ritchie. *The C Programming Language.* Prentice Hall, 1978.

[Khac79] L.G. Khachian. *A polynomial Algorithm for Linear Programming.* Doklady Akad. Nauk. USSR, 244(5):1093-1096, 1979.

[KnMa96] P.V. Knudsen and J. Madsen. *PACE: A Dynamic Programming Algorithm for Hardware/Software Partitioning.* International Workshop on Hardware/Software Codesign, pp. 85–92, 1996.

[Knud95] P.V. Knudsen. *Fine Grain Partitioning in Codesign.* Master Thesis, Department of Computer Science, Technical University of Denmark , 1995.

[Kone95] D. Konermann. *Integration von Steuerwerksynthese und Netzlistenerzeugung in das OSCAR Mikroarchitektur-Synthesesystem.* Diploma Thesis, University of Dortmund, 1995.

[KoRa94] Konstantinides and Rasure. *The Khoros Software Development Environment For Image And Signal Processing.* IEEE Transactions on Image Processing, VOL. 3, No. 3, pp. 243–252, May 1994.

[Koza92] J.R. Koza. *Genetic Programming: on the programming of computers by means of natural selection.* MIT Press, Cambridge, Massachusetts, 1992.

[KuMi88] D.C. Ku and G. De Micheli. *HardwareC - a language for hardware design.* Stanford University, Technical Report CSL-TR-90-419, 1988.

[LaMa93] B. Landwehr and P. Marwedel. *Intelligent Library Component Selection and Management in an IP-model based High-Level Synthesis System.* In Proceedings of the European Design Automation Conference (EURO-DAC), pp. 90–95, 1994.

[LaMa97] B. Landwehr and P. Marwedel. *A New Optimization Technique for Improving Resource Exploitation and Critical Path Minimization.* In Proceedings of the International Symposium on System Synthesis (ISSS), pp. 65–72, 1997.

[Land98] B. Landwehr. *ILP-basierte Mikroarchitektur-Synthese mit komplexen Bausteinbibliotheken.* Dissertation, University of Dortmund, 1998.

[Leup97] R. Leupers. *Retargetable Code Generation for Digital Signal Processors.* Kluwer Academic Publishers, 1997.

[Levi96] D. Levine. *Users Guide to the PGA-Pack Parallel Genetic Algorithm Library.* `ftp://ftp.mcs.anl.gov/pub/pgapack/user_guide.ps` Mathematics and Computer Science Division, Argonne National Laboratory, 1996.

[Liem97] C. Liem. *Retargetable Compilers for Embedded Core Processors.* Kluwer Academic Publishers, 1997.

[LiVe94] B. Lin and S. Vercauteren. *Synthesis of Concurrent System Interface Modules with Automatic Protocol Conversion Generation.* In Proceedings of the International Conference on Computer-Aided Design (ICCAD), pp. 101–108, 1994.

[LMD94] B. Landwehr, P. Marwedel and R. Dömer. *OSCAR: Optimum Simultaneous Scheduling, Allocation and Resource Binding Based on Integer Programming.* IFIP Workshop on Logic and Architecture Synthesis, pp. 381–400, 1993.

[LMM97] B. Landwehr, P. Marwedel and I. Markhof. *Exploiting Isomorphism for Speeding-Up Instance-Binding in an Integrated Scheduling, Allocation and Assignment Approach to Architectural Synthesis.* In Proceedings of the Conference on Computer Hardware Description Languages and their Application, 1997.

[Lore97] M. Lorenz. *Mikroarchitektur-Synthese mit genetischen Algorithmen.* Diploma Thesis, University of Dortmund, 1997.

[LPV94] Y.-T. Lai, M. Pedram and S.B.K. Vrudhula. *EVBDD-based Algorithms for Integer Linear Programming, Spectral Transformation and Function Decomposition.* IEEE Transactions on Computer Aided Design of Integrated Circuits and Systems, Vol 13, No. 8, pp. 959–975, 1994.

[LVM96a] B. Lin, S. Vercauteren and H. De Man. *Embedded Architecture Co-Synthesis and System Integration.* International Workshop on Hardware/Software Codesign, pp. 2-9, 1996.

[LVM96b] B. Lin, S. Vercauteren and H. De Man. *Constructing Application-Specific Heterogeneous Embedded Architectures for Custom HW/SW Applications.* In Proceedings of the Design Automation Conference (DAC), pp. 521–526, 1996.

[Madi96] V.K. Madisetti. *Rapid Digital System Prototyping: Current Practice, Future Challenge.* IEEE Design and Test of Computers, pp. 12-31, Fall 1996.

[MaGo95] P. Marwedel and G. Goossens. *Code Generation for Embedded Processors.* Kluwer Academic Publishers, 1995.

[Marm93] P. Marwedel, B. Landwehr, I. Markhof et al. *Endbericht der Projektgruppe MARMOR.* Internal Report, University of Dortmund, 1993.

[MBLR95+] H. De Man, I. Bolsens, B. Lin, K. Van Rompaey, S. Vercauteren and D. Verkest. *Co-Design of DSP Systems.* NATO Advanced Study Institute Workshop on Hardware/Software Codesign, Lake Como, Italy, June 1995.

[MeFa76] P. Merlin and D.J. Faber. *Recoverability of Communication Protocols.* IEEE Transactions on Communications, Vol. Com-24, No. 9, 1976.

[MeNa97] K. Mehlhorn and S. Näher. *The LEDA Platform of Combinatorial and Geometric Computing.* Cambridge University Press, 1997.

[MGB93] R.S. Mitra, B. Guha and A. Basu. *Rapid Prototyping of Microprocessor Based Systems.* In Proceedings of the International Conference on Computer-Aided Design (ICCAD)93, pp. 600–603, 1993.

[MGK+97] J. Madsen, J. Grode, P.V. Knudsen, M.E. Petersen and A. Haxthausen . *LYCOS: the Lyngby Co-Synthesis System.* Design Automation for Embedded Systems, vol. 2, no. 2, pp. 195–235, 1997.

[Mitr73] G. Mitra. *Investigation and some Branch and Bound Strategies for the Solution of Mixed Integer Linear Programs.* Math. Programming 4, pp. 155–170, 1973.

[Moto94] Motorola, Inc.. *DSP56001 Digital Signal Processor, User's Manual.* 1994.

[MQB95] R.S. Mitra, M.G. Qadir and A. Basu. *A Consistent Labelling Approach to Hardware Software Partitioning.* In Proceedings of the International Conference on VLSI Design95, pp. 19–24, 1995.

[MRB96a] R.S. Mitra, P.S. Roop and A. Basu. *An Overview of MICKEY: An Expert System for Automating the Design of Microprocessor Based Systems.* , SADHANA, Journal of the Indian Academy of Science, Vol.21, Pt.6, pp. 719-739, 1996

[MRB96b] R.S. Mitra, P.S. Roop and A. Basu. *A New Algorithm for Implementation of Design Functions by Available Devices.* IEEE Transactions on VLSI Systems, Vol.4, No.2, pp. 170–180, 1996.

[MSM97] V. Mooney, T. Sakamoto and G. De Micheli. *Run-Time Scheduler Synthesis For Hardware-Software Systems and Application to Robot Control Design.* International Workshop on Hardware/Software Codesign, pp. 95–99, 1997.

[NaGa94a] S. Narayan and D. Gajski. *Synthesis of System-Level Bus Interfaces.* In Proceedings of the European Conference on Design Automation (EDAC), pp. 395–399, 1994.

[NaGa94b] S. Narayan and D. Gajski. *Protocol Generation for Communication Channels.* In Proceedings of the Design Automation Conference (DAC), pp. 547–551, 1994.

[NaGa95] S. Narayan and D. Gajski. *Interfacing Incompatible Protocols Using Interface Process Generation.* In Proceedings of the Design Automation Conference (DAC), pp. 468–473, 1995.

[Nähe95] S. Näher. *The LEDA User Manual - Version 3.1.* Max-Planck-Institut für Informatik, Saarbrücken, 1995.

[Nava93] Z. Navabi. *VHDL - Analysis and Modelling of Digital Systems.* Mc Graw Hill, 1993.

[Nemh88] G. Nemhauser. *Integer and combinatorial optimization.* Wiley-Interscience series in discrete mathematics and optimization, 1988.

[Neum75] K. Neumann. *Operations Research Verfahren.* Band 1, Carl Hanser Verlag, 1975.

[NiMa95] R. Niemann and P. Marwedel. *Hardware/Software Partitioning using Integer Programming.* In Proceedings of the European Design & Test Conference (ED&TC), pp. 473–479, 1996.

[NiMa97] R. Niemann and P. Marwedel. *An Algorithm for Hardware/Software Partitioning using Mixed Integer Linear Programming.* Design Automation for Embedded Systems, vol. 2, no. 2, pp. 165–193, 1997.

[NiMa98] R. Niemann and P. Marwedel. *Synthesis of Communicating Controllers for Concurrent Hardware/Software Systems.* In Proceedings of the Design Automation and Test in Europe (DATE), 1998.

[NVG92] S. Narayan, F. Vahid and D. Gajski. *System specification with the SpecCharts language.* IEEE Design and Test of Computers, Dec. 1992.

[OrBo97] R.B. Ortega and G. Borriello. *Communication Synthesis for Embedded Systems with Global Considerations.* International Workshop on Hardware/Software Codesign, pp. 69–73, 1997.

[OSL92] IBM Corp. *Optimization Subroutine Library (OSL), Guide and Reference.* `http://www.research.ibm.com/osl/`, 1992.

[Papa82] C.H. Papadimitriou. *Combinatorial Optimization. Algorithms and Complexity.* Prentice Hall, 1992.

[PeKu93] Z. Peng and K. Kuchcinski. *An Algorithm for Partitioning of Application Specific Systems.* In Proceedings of the European Conference on Design Automation (EDAC), pp. 316–321, 1993.

[Petr62] C.A. Petri. *Kommunikation mit Automaten.* Dissertation, Bonn, 1962.

[PG245] Projektgruppe 245. *Endbericht der Projektgruppe 245 – Digitale Audiosignalverarbeitung mit FPGAs.* Forschungsbericht, University of Dortmund, 1997.

[PG293a] Projektgruppe 293. *Zwischenbericht der Projektgruppe 293 – Entwurf und Realisierung eines Fuzzy-Coprozessors auf der Basis von FPGAs und digitalen Signalprozessoren.* Forschungsbericht, University of Dortmund, 1997.

[PG293b] Projektgruppe 293. *Endbericht der Projektgruppe 293 – Entwurf und Realisierung eines Fuzzy-Coprozessors auf der Basis von FPGAs und digitalen Signalprozessoren.* Forschungsbericht, University of Dortmund, 1997.

[PGL+97] P. Paulin, G. Goossens and C. Liem. *Embedded Software in Real-time Signal Processing Systems: Application and Architecture Trends.* IEEE (Special Issue on HW/SW Co-design), 1997.

[Reis85] W. Reisig. *Petri-Netze - Eine Einführung.* Studienreihe Informatik, Springer Verlag Berlin, 1985.

[RVBM96] K. Van Rompaey, D. Verkest, I. Bolsens and H. De Man. *CoWare - A Design Environment for Heterogeneous Hardware/Software Systems.* In Proceedings of the European Design Automation Conference (EURO-DAC), 1996.

[Schw75] H.P. Schwefel. *Evolutionsstrategie und numerische Optimierung.* Dissertation, Technical University of Berlin, 1975.

[Smar96] J. Smart. *Reference Manual for wxWindows 1.66: a portable C++ GUI toolkit.* Artificial Intelligence Application Institute, University of Edinburgh, 1996.

[SMB97] D. Saha, R.S. Mitra and A. Basu. *Hardware Software Partitioning using Genetic Algorithms.* In Proceedings of the International Conference on VLSI Design97, pp. 155-160, 1997.

[SrBr91] M.B. Srivastava and R.W. Brodersen *Rapid-Prototyping of Hardware and Software in a Unified Framework.* In Proceedings of the International Conference on Computer-Aided Design (ICCAD), pp. 152–155, 1991.

[StSh86] L. Sterling and E. Shapiro. *The Art of Prolog.* MIT Press, Cambridge, Massachusetts, 1996.

[StWo97] J. Staunstrup and W. Wolf. *Hardware/Software Co-Design: Principles and Practice.* Kluwer Academic Publishers, 1997.

[Syno92] Synopsys. *Reference Manual Version 3.0.* 1992.

[TBT97] J. Teich, T. Blickle and L. Thiele. *An Evolutionary Approach to System-Level Synthesis.* International Workshop on Hardware/Software Codesign, pp. 167–171, 1997.

[Teic97] J. Teich. *Digitale Hardware/Software Systeme.* Springer Verlag Berlin, 1997.

[ThMo91] D.E. Thomas and P. Moorby. *The Verilog Hardware Description Language.* Kluwer Academic Publishers, 1991.

[UpKo94] B.P. Upender and P.J. Koopman Jr.. *Communication Protocols for Embedded Systems.* Embedded Systems Programming, 7(11), pp. 46–58, November 1994.

[Vahi97] F. Vahid. *Modifying Min-Cut for Hardware and Software Functional Partitioning.* International Workshop on Hardware/Software Codesign, pp. 43–48, 1997.

[VaLe97] F. Vahid and T.D. Le. *Extending the Kernighan/Lin Heuristic for Hardware and Software Functional Partitioning.* Design Automation for Embedded Systems, vol. 2, no. 2, pp. 237–261, 1997.

[Vant94] Vantage Analysis Systems. *Vantage Optium Online Manual, V5.1.* 1994.

[VaTa97] F. Vahid and L. Tauro. *An Object-Oriented Communication Library for Hardware-Software Codesign.* International Workshop on Hardware/Software Codesign, pp. 81–86, 1997.

[VGG94] F. Vahid, J. Gong and D. Gajski. *A Binary-Constraint Search Algorithm for Minimizing Hardware during Hardware/Software Par-*

titioning. In Proceedings of the European Design Automation Conference (EURO-DAC), pp. 214–219, 1994.

[VLM96] S. Vercauteren, B. Lin and H. De Man. *A Strategy for Real-Time Kernel Support in Application-Specific HW/SW Embedded Architectures.* In Proceedings of the Design Automation Conference (DAC), June 1996.

[VNG91] F. Vahid, S. Narayan and D.D. Gajski. *SpecCharts: A language for system level synthesis.* In Proceedings of the International Symposium on Computer Hardware Description Languages and their Applications, 1991.

[WaBo94] E.A. Walkup and G. Boriello. *Interface timing verification with application to synthesis.* In Proceedings of the Design Automation Conference (DAC), 1994.

[Wege93] I. Wegener. *Spezialvorlesung Operational Research.* Skript zur Vorlesung, University of Dortmund, 1993.

[WiCa97] J. Wilberg and R. Camposano. *VLIW Processor Codesign for Video Processing.* Design Automation for Embedded Systems, vol. 2, no. 1, pp. 79–119, 1997.

[Wirt71] N. Wirth. *The Programming Language Pascal.* Acta Informatica, Vol. 1, No. 1, 1971, pp. 35–63, Springer-Verlag.

[Xili94] Xilinx. *XACT Hardware & Peripherals Guide.* 1994.

[YeWo95] T.-Y. Yen and W. Wolf. *Communication Synthesis for Distributed Embedded Systems.* In Proceedings of the International Conference on Computer-Aided Design (ICCAD), 1995.

[ZDG97] J. Zhu, R. Dömer, D.D. Gajski *Syntax and Semantics of the SpecC Language.* Proceedings of the Synthesis and System Integration of Mixed Technologies, 1997.

[ZVSM94] V. Zivojnovic, J.M. Velarde, C. Schläger and H. Meyr. *A DSP-oriented Benchmarking Methodology.* In Proceedings of the International Conference on Signal Processing Applications and Technology (ICSPAT), 1994.

Notations

Variable	Def.	Description
a_i	14	i-th allele of chromosome
b_k	2	k-th communication channel
b_{i_1,i_2}	13	decision variable for scheduling v_{i_1} and v_{i_2}
C	1	set of system components
C^a	12	hardware area of the complete design
C_k^a	12	hardware area required on h_k
$c_{i,k}^a$	12	hardware area required by v_i on h_k
C^{dm}	12	software data memory of the complete design
C_k^{dm}	12	software data memory required on p_k
$c_{i,k}^{dm}$	12	software data memory required by v_i on p_k
c_l	1	system component
C^{pm}	12	software program memory of the complete design
C_k^{pm}	12	software program memory required on p_k
$c_{i,k}^{pm}$	12	software program memory required by v_i on p_k
$c_{i,k}^{th}$	12	hardware execution time required by v_i on h_k
$c_{i,k}^{tr}$	12	reading communication time required by v_i on channel b_k
$c_{i,k}^{ts}$	12	software execution time required by v_i on p_k
$c_{i,k}^{tw}$	12	writing communication time required by v_i on channel b_k
C^t	12	total execution time
d_i	14	DONE-state for node v_i
E	1	set of edges of the partitioning graph
$\mathcal{E}$	2	set of communication channels
g_i	14	i-th gene of chromosome
G^P	1	partitioning graph
G^S	1	scheduling graph (=reduced partitioning graph)
$\mathcal{H}$	2	set of hardware component
$h_{j,l,k}$	3	j-th hardware instance of c_l on h_k
h_k	2	k-th hardware component
I	1	instantiation function
i	-	index for nodes
j	-	index for a the j-th hardware instance $h_{j,l,k}$
k	-	index for target architecture components
l	-	index for system components
$\mathcal{M}$	2	set of memories
$Map_{\mathcal{E}}$	7	set of allocated communication channels
$Map_{\mathcal{V}}$	7	set of allocated hardware and software implementations
MAX^a	12	design constraint for hardware area
MAX_k^a	12	resource constraint of hardware area on h_k
MAX^{dm}	12	design constraint for software data memory
MAX_k^{dm}	12	resource constraint of software data memory on p_k
MAX^{pm}	12	design constraint for software program memory
MAX_k^{pm}	12	resource constraint of software program memory on p_k
MAX^t	12	design constraint for total execution time
m_k	2	k-th memory of the target architecture

Variable	Def.	Description
$nx_{j,l,k}$	13	decision variable for hardware instance $h_{j,k,l}$
$NX_{l,k}$	13	variable for the number of hardware instances of c_l on h_k
$NY_{l,k}$	13	decision variable for software function $p_{l,k}$ on p_k
$\mathcal{P}$	2	set of processors
$\mathcal{PU}$	2	set of processing units
p_k	2	k-th processor of the target architecture
$p_{l,k}$	3	software function c_l on processor p_k
R	14	global RESET-state
r_i	14	RESET-state for node v_i
$r_{j,k,l}$	14	RESET-state for hardware instance $h_{j,l,k}$
r_k	14	RESET-state for processor p_k
S	14	global START-state
$\mathcal{T}$	2	target technology
T_i^D	12	execution time of node v_i
T_i^E	12	ending time of node v_i
T_i^S	12	starting time of node v_i
t_k	2	target technology component
V	1	set of nodes
$\mathcal{V}$	2	set of target technology components
V^I	1	set of computation nodes
V^R	1	set of nodes representing the READ-phase for wires
V^{RI}	1	set of nodes representing the READ-phase for inputs
V^{RW}	1	set of communication nodes
V^W	1	set of nodes representing the WRITE-phase for wires
V^{WO}	1	set of nodes representing the WRITE-phase for outputs
X	14	global DONE-state
x_i	14	START-state for node v_i
$x_{i,j,k}$	13	variable for mapping computation node v_i to $h_{j,k,l}$ on h_k
$X_{i,k}$	13	variable for mapping computation node v_i to ASIC h_k
$Y_{i,k}$	13	variable for mapping computation node v_i to processor p_k
$z_{i,k}$	13	variable for mapping communication node v_i to channel b_k
Z_i	13	variable for communication node v_i

Abbreviations

ABS	Antilock Brake System
A/D	Analog / Digital
ALU	Arithmetic Logical Unit
ALAP	As Last As Possible
ANSI	American National Standards Institute
ASAP	As Soon As Possible
ASIC	Application Specific Integrated Circuit
ASIP	Application Specific Instruction-set Processor
AT	Area/Time
ATM	Asynchronous Transfer Mode
BDD	Binary Decision Diagram
BSS	Braunschweig Synthesis System
CAD	Computer-Aided Design
CAN	Controller Area Network
CASTLE	Codesign and Synthesis Tool Environment
CDFG	Control Data Flow Graph
CFG	Control Flow Graph
CLB	Configurable Logic Block
CLP	Consistent Labelling Problem
COOL	CO-design toOL
COSYMA	Co-synthesis for Embedded Architectures
CSP	Communicating Sequential Processes
D/A	Digital / Analog
DECT	Digital European Cordless Telephone
DFG	Data Flow Graph
DFS	Depth First Search
DLS	Design Library System
DSP	Digital Signal Processor / Processing
DVB	Digital Video Broadcast
DVD	Digital Video Disk
EA	Evolutionary Algorithm
ECAD	Electronic Computer-Aided Design
EP	Evolutionary Programming
ES	Evolutionary Strategy
FFT	Fast Fourier Transformation
FIFO	First In First Out
FPGA	Field Programmable Gate Array
FSM	Finite State Machine
GA	Genetic Algorithm
GP	Genetic Programming
GSM	Global System for Mobile Communication
HCFSM	Hierarchical Concurrent Finite State Machine
HDL	Hardware Description Language

HDTV	High-Definition TV
HFSM	Hierarchical Finite State Machine
IEEE	Institute of Electrical and Electronics Engineering
ILP	Integer Linear Programming
IP	Integer Programming
I/O	Input/Output
ISDN	Integrated Services Digital Network
ISO	International Standard Organization
ITU	International Telecommunication Union
JPEG	Joint Photographic Expert Group
LEDA	Library of Efficient Datatypes and Algorithms
LOTOS	Language of Temporal Ordering Specification
LP	Linear Programming
LSB	Least Significant Bit
LYCOS	Lyngby Cosynthesis
MILP	Mixed Integer Linear Programming
MPEG	Motion Picture Expert Group
MSB	Most Significant Bit
OSBACK	OScar BACKend
OSCAR	Optimum Simultaneous sCheduling, Allocation and Resource binding
OSI	Open Systems International
OSL	Optimization Subroutine Library
PCI	Peripheral Component Interconnect
PSM	Program State Machine
RAM	Random Access Memory
RISC	Reduced Instruction-Set Computers
ROM	Read Only Memory
RPC	Remote Procedure Call
RT	Register Transfer
SDL	Specification Description Language
SIR	System Intermediate Representation
SPI	Software Procedural Interface
TOSCA	TOols for System Codesign Automation
USB	Universal Serial Bus
VHDL	VHSIC Hardware Description Language
VHSIC	Very High Speed Integrated Circuit
VLSI	Very Large Scale Integration
VTIP	VHDL Tool Integration Platform

Index

A
A/D, 8
ABS, 4
address bus, 173, 174
ADEPT, 20
ALAP, 93, 107
algebraic transformation, 68
algorithm
    binary-constraint search, 54
    clustering, 55
    evolutionary, 83
    genetic, 48, 55, 85, 115, 195
    greedy, 48
    heuristic, 48
    probabilistic, 48
    simplex, 81
algorithmic level, 2, 3
allele, 84
allocation, 3, 20, 67, 143
ALU, 2
ANSI-C, 32, 38, 62, 64
application domain, 48
arbitration scheme, 142
    dynamic-priority, 142
    fixed-priority, 142, 153
architectural template, 143
architecture
    multi-processor, 18
    single-processor, 15
    target, 22, 44, 47, 55
        heterogeneous, 174, 193
architecture body, 40
array, 170
ASAP, 93, 107
ASIC, 2, 6, 7, 75
ASIP, 6, 62, 197
assembler program, 64
AT-curve, 195
ATM, 7
audio, 7
automation, 48
availability, 35

B
basic block, 30, 68
BDD, 82
behavioral completion, 34, 41
behavioral description, 40
behavioral domain, 1
bi-directional signal, 146, 154
big endian, 139, 167
binding, 3, 67
bit, 43
block, 2
block diagrams, 31
board, 2, 8
bottom-up, 34, 41
branch-and-bound method, 82
broadcast mechanism, 141
BSS, 16
bus, 2, 8, 141
bus arbiter, 24, 142, 145, 153
bus driver, 154
bus generation, 142, 143
bus grant, 153, 173
bus request, 153, 173
bus snooper, 149, 153

C
C, 15–17, 19, 27, 38, 62, 64, 115, 144, 156, 165, 166, 168, 171, 178, 183, 184, 194
C++, 14, 15, 27, 183, 184
C-COMPILER, 185
CAN, 142
capacitor, 2
CASTLE, 14, 15
CHINOOK, 18, 143
chip, 2, 8

chromosome, 84
circuit level, 2, 3
CLB, 78
CLOCK, 68
CLP, 17
co-processor , 15
co-simulation, 9, 197
co-synthesis, 9, 133, 193, 195, 196
co-validation, 10
co-verification, 10
code quality, 10
CODES, 20
CODESIGN, 20
communication, 34, 41, 52, 140
  blocking, 137, 140
  inter-process, 4
  message passing, 34, 59, 140, 144
    blocking, 59
    non-blocking, 59
  non-blocking, 140
  shared memory, 34, 59, 70, 75, 140, 141, 144
communication channel, 59, 75, 140
communication protocol, 142
communication time
  reading, 77
  writing, 77
COMPASS, 68, 69
compilation, 137
compilation time, 10
compiler, 48
computation time, 48
computer-aided design, 1
concurrency, 34, 41
constraint, 45, 81, 91
  binding, 45
  design, 93
  mapping, 45, 91
  maximum timing, 45
  minimum timing, 45
  resource, 45, 92
  timing, 35, 41, 93
constraint violation, 120
control flow dominated, 5
control step, 68
control step list, 69
controller, 2
COOL, xvi, xvii, 11, 14, 22–25, 28, 41–45, 48, 49, 55–64, 70, 86, 96, 101, 110, 115, 128, 133, 134, 144–148, 151, 153, 155–157, 164–168, 174–176, 179, 180, 183–188, 190, 193–197
COSMOS, 18, 19, 143
cost, 74
cost estimation, 8, 62, 194, 196
  hardware, 67
  interface, 70
  software, 64
cost function, 48
cost metric, 48, 77, 88
  hardware, 64
  interface, 64
  software, 64
COSYMA, 15–18, 53
COWARE, 18, 19, 143
critical path, 64
crossover, 85
crossover probability, 117
crossover rate, 123
crossover type, 116, 123
CSP, 38
cutting-plane method, 82
cycle time, 68

D

D/A, 8
data abstraction, 33
data bus, 173, 174
data flow, 43
  dynamic, 14
  synchronous, 14
data flow dominated, 6
data memory, 64, 77–79
data path, 3, 144
data path controller, 3, 144, 171
data strobe, 173
data type conversion, 139
DAVIS, 176, 179
dead-lock, 10, 35
debugging, 5
DECT, 6
dedicated lines, 141
defuzzification, 174
design, 78
  system-level, 3
design constraint, 22, 44, 55, 74, 79, 88
design cost, 5, 79
design entity, 40
design environment, 44
design methodology management, 11, 195
design metric, 88
design quality, 78
design space exploration, 11, 16, 48, 86, 127, 137
DFL, 19, 62
DFS, 107
digital signal processing, 6, 30, 39
digital video broadcast (DVB), 6
digital video disks (DVD), 6
discrete event, 14
DLS, 184
Dolby AC3, 6
domain, 14
DSP, 6, 7

dynamic analysis, 62
dynamic programming, 16, 54, 55
dynamical profiling, 15

E

ellipsoid method, 82
embedded controller, 5
embedded processor, 7
emulation, 10
ending time, 88
entity declaration, 40
entity-relationship diagram, 31
environment characteristics, 35
Estelle, 36
Esterel, 19, 37
estimation method, 48
estimation tools, 48
evolutionary algorithm, 83
evolutionary approach, 54
evolutionary operator, 84
evolutionary programming, 85
evolutionary strategy, 85
exception, 34
execution order, 61, 73
execution time, 88
  average, 62
  hardware, 70, 77
  software, 77
  software , 64
  total, 53
  worst-case, 62, 64
extended finite-state machine, 36

F

FIFO, 59
FIFO buffer, 141
finite-state machine, 2, 29
  co-design (CFSM), 19
  hierarchical concurrent (HCFSM), 29
fitness, 83
fitness function, 83, 116, 119
fixed-point, 43
flip-flop, 2
for, 43
form factor, 5
formal analysis, 35
formal verification, 10
FPGA, 10, 78
functional pipelining, 50, 52, 58
functional timing, 35, 41
functional verification, 137
fuzzification, 174

G

galaxy, 14
gate level, 2
gate netlist, 31
gatelevel, 3
GCLP, 54
gene, 84
generation, 83, 116
genetic parameters, 115
glue logic, 142
granularity, 48
granule, 8
graph
  control flow, 30
  control/data flow, 32, 68
  data flow, 30
  extended syntax, 16
  partitioning, 55, 71, 74
    colored, 56
  state-transition, 144, 157, 158
GSM, 6

H

hardware/software co-design, 8, 193
hardware/software mapping, 49, 50, 58, 79, 80
hardware/software partitioning, 9, 47, 193, 194, 196
  hardware-oriented, 53
  software-oriented, 53
hardware area, 70, 77–79
hardware component, 75
hardware implementation, 76
hardware instance, 76
hardware resource instance, 68
hardware sharing, 49, 51, 58, 94
HardwareC, 17, 38
heterogeneous modelling, 27
hierarchy, 34
  behavioral, 34, 41
  structural, 34, 41
high-definition TV (HDTV), 6
homogeneous modelling, 27

I

I/O, 8
I/O controller, 24, 145, 151
I/O port allocation, 143
individual, 83
inference, 174
information hiding, 33
initial population, 83
inner loop partitioning, 53
input handler, 152, 153
instance, 41
instruction cycle, 65, 66
instruction cycle time, 66
instruction set description, 64
integer, 43
integer linear programming (ILP), 48, 68
interfacing, 50, 58, 95

interior-point method, 82
interrupt, 138
interrupt service routine, 138
ISDN, 7

J

JPEG, 6

K

Kernighan/Lin heuristic, 54
KHOROS, 14

L

language
    applicative, 39
    data flow, 39
    declarative, 32
    formal description, 36
    hardware description, 37
    imperative, 32, 41
    parallel programming, 38
    programming, 32, 38
        synchronous, 37
    real-time system, 37
latency, 53, 88
LEDA, 183
library
    component, 62, 68
    cost, 22, 55, 63
    design, 57
    function, 68
    subroutine, 65
    system, 22, 41, 55
    target architecture, 22, 62, 70
    transformation, 68
library binding, 3
library classes, 41
library management, 168
linear programming (LP), 81
linguistic term, 174
linguistic variable, 174
LINUX, 183, 184
little endian, 139, 167
local extrema, 86
local memory, 155
logic analyzer, 179
logic level, 2, 3
logic-level minimization, 3
LOTOS, 35, 36
low power, 196
LYCOS, 15, 16, 54

M

maintenance, 5
manufacturing cost, 5, 8
mapping function, 80
mask, 2
Mealy-automaton, 29
memory, 2, 75, 155
memory address translation, 139
memory allocation, 135
memory mapped I/O, 16, 143, 163
MICKEY, 17, 144
microcontroller, 5
MILP, 55, 106
MINNIE, 17
mixed integer linear programming (MILP), 57
model, 27
    activity-oriented, 30
    data-oriented, 31
    dynamic data flow, 6
    heterogeneous, 32
    object-oriented, 33
    state-oriented, 28
    structure-oriented, 30
    synchronous data flow, 6
model executability, 35
Moore-automaton, 29
MOTIF, 184
MOTOROLA, 65, 75, 146, 169, 176, 178, 185
MPEG, 6
MS WINDOWS, 184
MSCE, 20
multi-way channel, 141
multiplexer, 2
mutation, 85
mutation probability, 117
mutation rate, 123
mutation type, 117

N

node
    communication, 71, 117
    computation, 71, 117
    decision, 30
    dominator, 104
non-determinism, 35, 63
non-preemptive, 61

O

objective function, 81, 91
Occam, 17, 38
one-point, 123
OOCL, 144
operation, 68
operation replacement, 139
operations research, 81
optimization, 3
optimization aspects, 48
optimization goal, 48
optimization problem
    linear, 86
    non-linear, 86
optimization technique, 48

OsBack, 67–70, 146
Oscar, 24, 62, 67–69, 128, 146, 154, 165–167, 170, 184, 185, 194
Osl, 69, 127, 185
outer loop partitioning, 53
output handler, 152

P

PACE, 54
Partif, 19
partitioning, 20
partitioning quality, 48
`Pascal`, 32
PCI, 142
perfect synchrony hypothesis, 37
performance, 5, 8
Petri net, 28
  high-level, 37
  high-level time, 37
  time, 37
PgaPack, 115, 117, 184
physical design, 3
physical domain, 1
physical level, 1, 2
placement, 3
point-to-point connection, 141
Polis, 18–21
polygon, 2
population, 83
population replacement size, 116
population size, 116
power consumption, 5, 8
predecessor, 103
primitive function, 17
problem
  binary programming (0-1-IP), 83
  constraint-satisfaction, 106
  contiguous, 82
  discrete, 82
  extended partitioning, 54, 128, 195
  hardware/software partitioning, 80
  implementation selection, 54
  integer linear programming, 82
  linear programming, 81
  mapping, 101
  mixed integer linear programming (MILP), 83, 195
  optimization, 106
  partitioning, 47, 49
  scheduling, 101
process algebra, 17, 36, 38
processing unit, 75
processor, 2, 75
processor cores, 8
profiling, 16, 62
program memory, 64, 77–79
program-state machine, 20, 33, 38
programmability, 5
programming construct, 34, 41
`Prolog`, 32
protocol
  2-phase handshake, 142
  4-phase handshake, 142
  communication, 142
  handshake, 142, 172
  synchronous wait, 143
protocol conversion, 142, 197
protocol generation, 142, 143
protocol selection, 134
Ptolemy, 14

Q

Quenya, 16

R

RAM, 7
re-partitioning, 10
reactive, 4
`READ`, 59
read/write, 173
real-time constraint, 4
recombination, 85
Record, 62, 197
refinement, 3, 20
  control-related, 137
  data-related, 139
  hardware, 170
  software, 167
  specification, 9, 133, 136
register, 2
register-transfer level, 2, 3
reliability, 5, 8
repair mechanism, 83, 116, 119, 129
reset, 68
resistor, 2
resource conflict, 52
resource constraint, 78, 88
resource cost, 78, 88
retargetable code generator, 62, 197
retargetable compiler, 10
RISC, 7, 15
ROM, 7
routing, 3
RPC, 19
RS-232, 142
RT netlist, 31
rule base, 174

S

safety, 5
safety-critical, 10
sample rate, 7
scheduler, 14, 120
  run-time, 24, 60, 138, 144, 148

scheduling, 3, 50, 52, 56, 58, 60, 67, 97, 103, 106, 122
  heuristic, 57, 101, 113, 195
  list, 102, 106
SDL, 18, 36
selection, 83
selection type, 116
servicing, 5
SIERA, 20
Silage, 39
simplex algorithm, 81
simulated annealing, 16, 48, 53, 55
simulation, 10, 137
single point crossover, 85
SLIF, 20
software compilation, 10
software function, 76
software implementation, 76
Solar, 19
SpecC, 38
SpecCharts, 20, 38
specification
  refined, 136
SPECSYN, 18, 20, 143, 144, 193
SpeX, 17
SPI, 184, 185
standardization, 36
star, 14
start/stop mechanism, 148
starting time, 88
state, 156
state encoding, 3
state minimization, 3
state-transition, 34
StateCharts, 37
StateCharts, 17, 20, 37
STATEMATE, 14
static analysis, 57, 62, 194
static profiling, 15
stimuli file, 44
stopping criterion, 83, 117
string, 84, 119
structural description, 40
structural domain, 2
SUNOS 4.1, 183, 184
SYMPHONY, 143
synchronization, 34, 41, 137
  message passing, 137, 171
  shared memory, 138, 164, 171
synchronization mechanisms, 137
SYNOPSYS, 24, 68, 69, 146, 154, 179, 184, 185
synthesis, 3
  behavioral, 3
  circuit-level, 3
  communication, 133, 134
  controller, 3
  hardware, 137
  high-level, 3, 6, 62, 67, 194
  interface, 140, 197
  logic-level, 3
synthesis specification file, 68
synthesis tool, 48
system
  embedded, 4, 193
  embedded data-processing, 6
  general-purpose, 4
  hardware/software, 7
  heterogeneous, 7
  informational, 31
  reactive, 5
  real-time, 6
  single-board, 8
  single-chip, 8
  special-purpose, 4
  transformational, 6, 30, 151
system component, 41, 71
system controller, 24, 144, 148, 171
system level, 1, 2
system specification, 8, 27, 194, 196

T

tabu search, 55
target technology, 74, 75
technology mapping, 3
telecommunication, 7
terminal-based, 4
testability, 196
testing, 5
time-to-market, 5
timing aspect, 35
top-down, 34, 41
topological sorting, 107
TOSCA, 16, 17
total execution time, 79
transformation rule, 68
transistor, 2
transistor level, 2
transputer, 38
two-point, 123

U

unbounded loop, 63
uniform, 123
UNIX, 183, 184
urgency metric, 174
USB, 142
user-interaction, 48

V

VANTAGE OPTIUM, 24, 44, 145, 156, 180, 184–187, 191
variable, 80
variable folding, 139
variation, 83

verifiability, 35
Verilog, 15, 18, 37
VHDL, 15–17, 19, 20, 22, 24, 27, 28, 36–38, 40, 41, 43, 44, 62, 68, 144–147, 154–156, 165, 166, 171, 172, 176, 178, 179, 184–187, 194, 196
video, 7
VLSI, 1
VTIP, 62, 184, 185
VULCAN, 16–18, 53

W

wait-state, 174
WINDOWS 3.1, 184
WINDOWS 95, 184
WINDOWS NT, 184
wire, 8
worm-hole mechanism, 14
WRITE, 59
wxWINDOWS, 183, 184

X

XILINX, 177, 179
XT, 184
XVIEW, 184

Y

Y-chart, 1

Z

Z-value, 146, 154